遗传学综合双语实验教程
第 2 版

李卫东　主　编

王沛政　李成奇　副主编

北京理工大学出版社
BEIJING INSTITUTE OF TECHNOLOGY PRESS

图书在版编目（CIP）数据

遗传学综合双语实验教程 / 李卫东主编. —2版. —北京：北京理工大学出版社，2019.1

ISBN 978-7-5682-6662-8

Ⅰ.①遗… Ⅱ.①李… Ⅲ.①遗传学–实验–双语教学–教材 Ⅳ.①Q3-33

中国版本图书馆CIP数据核字（2019）第013917号

出版发行 / 北京理工大学出版社有限责任公司
社　　址 / 北京市海淀区中关村南大街5号
邮　　编 / 100081
电　　话 / （010）68914775（总编室）
　　　　　（010）82562903（教材售后服务热线）
　　　　　（010）68948351（其他图书服务热线）
网　　址 / http：//www.bitpress.com.cn
经　　销 / 全国各地新华书店
印　　刷 / 涿州市新华印刷有限公司
开　　本 / 710毫米 × 1000毫米　1/16
印　　张 / 11　　　　　　　　　　　　　责任编辑 / 梁铜华
字　　数 / 179千字　　　　　　　　　　　文案编辑 / 梁铜华
版　　次 / 2019年1月第2版　2019年1月第1次印刷　　责任校对 / 杜　枝
定　　价 / 35.00元　　　　　　　　　　　责任印制 / 施胜娟

图书出现印装质量问题，请拨打售后服务热线，本社负责调换

内容简介：

本书涉及经典遗传学、细胞遗传学、分子遗传学、群体遗传学和数量遗传学等内容，合计 23 个实验，基本上体现了基础遗传学实验教学的需求。全书内容和结构编排新颖，实验设计有实验原理、实验记录、实验报告等基本要求；具体实验内容设计了实验原理、基本材料、实验步骤、结果辨析、注意事项等内容。这些是培养学生具备缜密思维和分析能力不可缺少的途径。

Brief Introduction：

This book relates to classical genetics, cytogenetics, molecular genetics, population genetics and quantitative genetics and other content and has a total of 23 experiments. Basically, It reflects the teaching demand of the fundamental genetics experiment. The content and structure arrangement was new, and the experimental design includes principles, test records, test reports and other basic requirements. The concrete experiment content designing has the experiment principles, basic materials, experimental steps, analyses of results, matters needing attention. These are the necessary approach of cultivating students with gentry densely thinking and analytical skills.

前　言

　　遗传学是介绍孟德尔遗传、染色体和分子基础的一门课程。实验室操作主要强调基本遗传技术动手操作经验，其技术包含一些生物遗传杂交的构建与分析、染色体材料的制备和分析，DNA 的检测以及通过凝胶电泳对 DNA 序列变化的分析。很多的实验材料都包括果蝇，果蝇是遗传学实验中经典的生物体。本书包括四部分共 23 个实验，每个实验的实验原理已构建好，可以让您熟悉遗传学这门课。

　　本书实验数目较多，还包括一些分子水平上的实验技术，而实验时间有限，不可能都做，所以可以根据具体情况，酌情选择。限于时间，有些实验未收集，有待以后补充。

　　此次编写是双语教材编写的尝试和探索，疏漏和错误在所难免，愿遗传学同道在格式、思路以及实验本身方面不断提出宝贵意见，以期再版时不断完善。

　　本书编写具体分工情况：李卫东老师（海南热带海洋学院）编写第一部分。王沛政老师（海南热带海洋学院）编写第二部分和第三部分，李成奇老师（河南科技学院）编写第四部分。

　　本书由海南热带海洋学院资助出版。

编　者

2018 年 11 月

Preface

Genetics is an introductory course that deals with the laws of Mendelian inheritance and their chromosomal and molecular basis. The major laboratory investigations emphasize hands-on experience with basic genetic techniques, including construction and analysis of genetic crosses with several organisms, preparation and analysis of chromosome material, examination of DNA and calculation and analysis of patterns of DNA sequence variation by electrophoresis. Many of the exercises involve Drosophila melanogaster, the classic organism of experimental genetics. This book includes four aspects and a total of 23 experiments, of which the principles have been established to acquaint you with Genetics.

The number of experiments in this book is large. Moreover, this book includes experimental techniques based on some molecular level. The time for doing experiments in class is limited, so all of the experiments can not be done. Therefore, you can make appropriate choice according to the specific circumstances. Due to the limitation of time, some experiments are not collected, which need to be added in future.

This book is an attempt and exploration of bilingual textbook, so omissions and mistakes can hardly be avoided. We welcome your comments and suggestions on the format, presentation of ideas, and experiments themselves in order to improve in the next edition.

In this book, Dr Li Weidong (Hainan Tropical Ocean University) wrote the first part, Dr Wang Peizheng (Hainan Tropical Ocean University) wrote the second and third part and Dr Li Chenqi(Henan Institute of Science and Technology) wrote the last part.

This book is sponsored by Hainan Tropical Ocean University.

Editor

2018.11

目　录

第一部分

实验一　洋葱（大蒜）根尖有丝分裂标本的观察

一、实验目的

（1）制作洋葱根尖细胞有丝分裂的标本，并能观察到有丝分裂的每个时期。
（2）更好地理解有丝分裂的过程和阶段。
（3）分析并估计有丝分裂每个阶段的相对时间。

二、实验原理

真核生物中，DNA 的复制通常伴随着有丝分裂的过程。有丝分裂保证了每个子细胞都得到一份复制的染色体。染色体在有丝分裂过程中经历以下几个阶段：前期、中期、后期和末期。细胞质的实际分裂叫作胞质分裂，发生在末期。胞质分裂之前的每个阶段及特定的事件有助于已经复制的染色体的有序分布。

三、实验材料

洋葱根尖、复式显微镜、洋葱载玻片样片、拨针、载玻片、盖玻片、镊子、滤纸、刀片、乙酸、地衣红、1 mol/L 盐酸、卡诺氏固定液（乙醇混合物∶冰醋酸 =3∶1 或 9∶1）、10% 的冰醋酸、蒸馏水或去离子水溶液以及卡宝品红。

四、实验步骤

（一）压片制作的准备程序

（1）用剪刀剪下两段长约 1 cm 的洋葱根尖，转移到离心管中。
（2）用卡诺氏固定液固定 12 h。

（3）倒掉卡诺氏固定液，往离心管中加入 2/3 体积的 1 mol/L 的 HCl。（注意：HCl 是强酸，小心操作）。

（4）将离心管置于 60℃ 水中水浴，让根尖孵育 12 min，然后移出离心管。

（5）用镊子小心地将根尖移到载玻片上。

（6）用滴瓶冲洗根尖 3 次。

（二）染色

（1）冲洗 3 次后，添加卡宝品红染液或乳酸—乙酸—地衣红染液。

（2）在染液中孵育 12 min。在此期间，根尖开始变红。

（3）倒掉染色液并冲洗根尖 3 次。

（三）制作根尖压片

（1）将根转移到一个洁净的载玻片上，滴一滴水。

（2）用刀片切掉根尖未染色部分并丢弃。

（3）用盖玻片盖住根尖，然后在盖玻片覆盖的一面小心地用解剖探针末端下压，用力但不要扭动或侧向推动。根尖应被展开至直径为 0.5~1 cm。

（4）在显微镜下检查有丝分裂的各个阶段。

五、结果

观察洋葱根尖压片，在 10 倍镜下寻找目标。寻找区域具有与细胞相比较大的细胞核。在这些细胞中将会发现各时期的有丝分裂。转换到 40 倍镜仔细观察。由于前期和前中期很难区分，因此将其划分为前期。

绘制你所观察到的图像，并说明是哪一个时期的有丝分裂。

问题：

（1）为什么要使用洋葱根观察有丝分裂？

①洋葱根很容易生长，并能很快大量增殖。

②根尖细胞分裂旺盛，便于查看各时期的有丝分裂。

③可以将染色体染色，使它们更容易被观察到。

（2）有丝分裂各阶段特点是什么？

前期：

中期：

后期：

末期：

（答案略。）

附录：

乳酸—乙酸—地衣红染液：一份 A 液 + 一份 B 液 + 一份水。

地衣红冰醋酸溶液（A）：约 2 g 地衣红 +100 mL 冰醋酸，加热（谨防沸腾），充分溶解后过滤。

地衣红乳酸溶液（B）：约 2 g 地衣红 +100 mL 乳酸，加热（谨防沸腾），充分溶解后过滤。

卡宝品红：约 0.3 g 地衣红 +90 mL 石碳酸酚（5%）+11 mL 冰醋酸 +11 mL 甲醛（37%），充分溶解。

Experiment 1　Looking at the Mitosis Using Onion （Garlic）Roots

Ⅰ. Experimental Objectives

(1) Prepare your own specimens of onion roots with which you can visualize all of the stages of the mitosis.

(2) Better understand the process and stages of the mitosis.

(3) Apply an analytical technique by which the relative length of each stage of the mitosis can be estimated.

Ⅱ. Experimental Principles

DNA replication in eukaryotes is accompanied by the process called mitosis which assures that each daughter cell receives one copy of the replicated chromosomes. During the process of mitosis, the chromosomes pass through several stages known as prophase, metaphase, anaphase and telophase. The actual division of the cytoplasm is called cytokinesis and occurs at telophase. During each of the stages prior to cytokinesis, particular events occur, which contributes to the orderly distribution of the replicated chromosomes.

Ⅲ. Experimental Materials

Onion roots, compound microscopes, prepared slides of longitudinal sections of onion roots, dissecting needles, slides, cover slips, forceps, filter paper, scalpel, acetic acid, acetic orcein, 1mol/L Hydrochloric acid, Carnoy's fixative (Mixture of ethyl alcohol: glacial acetic acid in proportions of 3 ∶ 1 or 9 ∶ 1), 10% solution of glacial acetic acid, distilled or deionized water, and carbol fuchsin.

Ⅳ. Experimental Procedures

Preparing root tip squashes:

(1) Using scissors, cut 2 root tips about 1 cm long, and transfer them into a plastic micro-tube.

(2) Treat them with Carnoy's fixative for 12 h.

(3) Drain the Carnoy's fixative and fill the centrifuge tube to about 2/3 full with 1 mol/L HCl (Caution: work with the HCl carefully. It is a strong acid).

(4) Place the centrifuge tube into a 60℃ water bath, and allow the roots to incubate for 12 minutes. After that remove the tube from the water bath.

(5) Carefully transfer the root tips to a small petri plate using forceps.

(6) Rinse the root tips 3 times with the water, using the dropper bottle.

Staining the chromosomes:

(1) Cover the root with the Carbol fuchsin stain/Lactic acid, acetic and orcein dye.

(2) Incubate the roots into the stain liquid for 12 minutes. During this process the very tip of the root will begin to turn red.

(3) Remove the stain and again rinse the roots 3 times with water.

Making the root tip squash:

(1) Transfer a root to a clean microscope slide and drip a drop of water on it.

(2) Using a razor blade, cut off the unstained part of the root, and discard it.

(3) Cover the root tip with a cover slip, and then carefully push down the cover slide with the wooden end of a dissecting probe. Push hard, but do not twist or push the cover slide sideways. The root tip should spread out to a diameter of 0.5–1cm.

(4) Examine the stages in mitosis.

V . Experimental Results

Observe the onion root tip squash under the 10 × objective. Look for the region that has larger nuclei relative to the size of the cell. The cells will display stages of mitosis. Switch to the 40 × objective to make closer observations. Prophase and prometaphase are difficult to distinguish, so all these cells are classified as prophase.

Record your observations in the table provided. Draw the image of what you have observed, and describe which stage it is in the mitosis.

Questions:

(1) Why do we use onion roots to view the mitosis?

① The onion root can grow easily and large numbers of roots will be got soon.

② The cells at the tip of the roots divide actively so it is easier to check the periods of mitosis.

③ Chromosomes can be stained to make them more easily observable.

(2) What is the distinguishing visible feature in each stage of mitosis?

Prophase:

Metaphase:

Anaphase:

Telophase:

(The answer is omitted.)

Appendix:

Lactic acid, acetic and orcein dye: an A solution, a certain quantity of B solution and a certain quantity of water.

Aceto-orcein stain (A): about 2 g orcein stain +100 mL glacial acetic acid, heating (not boiling), fully dissolving, then filtering.

Orcein-lactic acid stain (B): about 2 g orcein stain +100 mL lactic acid, heating (not boiling), fully dissolving, then filtering.

Carbol fuchsin: 0.3 g orcein stain + 90 mL carbolic acid (5%) +11 mL glacial acetic acid+11 mL formaldehyde (37%), fully dissolving.

实验二　减数分裂标本的制作与观察

一、实验目的

研究减数分裂对了解细胞的正常分裂是很必要的。在这个实验过程中，学生将识别雄性蝗虫不同阶段的减数分裂。将活体蝗虫解剖后，取出精腺并染色观察。

二、实验原理

减数分裂是子细胞染色体数目减半（从二倍体到单倍体）类型的细胞分裂，比如配子。减数分裂包含两个阶段：减数分裂 I 和减数分裂 II。减数分裂 I 时期同源染色体相互分离。减数分裂 II 时期染色体均等分离并形成 4 个子细胞。在减数分裂阶段，细胞学观察要准备的材料主要有精腺小管细胞或花蕾和花药中的花粉母细胞等。

三、实验材料

雄性蝗虫（无产卵器）、复式显微镜、蝗虫载玻片样片、拨针、载玻片、盖玻片、镊子、滤纸、刀片、50% 乙酸、2% 醋酸地衣红以及卡宝品红。

四、实验步骤

（1）洗手并戴上手套和护目镜。

（2）收集设备材料。

（3）准备工作区。

（4）从卡诺固定液中取出一个雄性蝗虫。

（5）取下附属肢体。

（6）沿着腹部做背侧切口。

（7）寻找鲜艳的橙色或黄色的块状物。用镊子将其小心取出（这些就是精腺）。

（8）把精腺放在含有少量蒸馏水的玻璃片上。用镊子将组织分成小片。

（9）取少量组织放在载玻片上。向载玻片上加 1 滴 50% 的乙酸。

（10）用滤纸吸干过多的乙酸，向玻片加 2 滴卡宝品红（或 2% 醋酸地衣红）。

（11）载玻片上放置盖玻片，用滤纸覆盖载玻片。

（12）用拇指按压盖玻片和滤纸（这将使精腺细胞核破裂）。

（13）把载玻片放在显微镜观察台上，观察各个时期的减数分裂。

（14）对工作台表面进行消毒并清洁工作区。摘除护目镜和手套并洗手。

五、实验结果

用准备好的睾丸和卵巢载玻片，分析并观察各阶段的减数分裂。描绘观察草图。

项目	视野 1	视野 2	视野 3	总数	百分比
分裂间期					
分裂前期					
分裂中期					
分裂后期					
分裂末期					

问题：

（1）列出有丝分裂和减数分裂的 3 个主要不同点。

在有丝分裂中，细胞核只分裂一次；而在减数分裂中，细胞核分裂两次。另一个不同点是有丝分裂产生两个子细胞，但是减数分裂产生高达 4 个子细胞。减数分裂过程中发生同源染色体的联会和非姐妹染色单体之间的交叉互换，而有丝分裂不发生。

（2）为什么减数分裂对有性生殖来说很重要？

减数分裂过程中，染色体数目减少到单倍体的数目可以用于受精。减数分裂还会发生非姐妹染色单体之间的交换，从而导致生物体的变异。

Experiment 2　Preparation and Observation of Meiosis Specimens

Ⅰ. Experimental Objectives

A study of meiosis is necessary for an understanding of normal cell division. In this laboratory exercise, students will identify the normal phases of meiosis in a male grasshopper. A live grasshopper is dissected, then the testes are removed and stained for observation.

Ⅱ. Experimental Principles

Meiosis is a type of cell division during which the number of chromosomes is halved (from diploid to haploid) in the daughter cells, for example, the gametes. The division includes two phases—meiosis I and meiosis II. Meiosis I is a reductional division in which the homologous chromosomes pairs separate from each other. Meiosis II is equational division and forms four daughter cells. Stages of meiosis can be observed in a cytological preparation of the cells of testis tubules or in the pollen mother cells of the anthers of flower buds.

Ⅲ. Experimental Materials

Male grasshoppers (no ovipositors), compound microscopes, prepared slides of grasshoppers, dissecting needles, slides and cover slips, forceps, filter paper, scalpels, 50% acetic acid, 2% acetic orcein, and carbol fuchsin.

Ⅳ. Experimental Procedures

(1) Wash hands and put on gloves and goggles.

(2) Assemble equipment and materials.

(3) Clean up work areas.

(4) Take out a male grasshopper from Carnoy's fluid.

(5) Remove the appendages.

(6) Make a dorsal incision along the abdomen.

(7) Look for brightly colored orange or yellow mass and carefully remove with forceps (These are the testes).

(8) Put the testes on a watch glass on which a small amount of distilled water is dripped. Divide the tissue into smaller pieces with forceps.

(9) Place a small amount of tissue on the slide. Drip a drop of 50% acetic acid to the slide.

(10) Dry with the filter paper. Drip 2 drops of carbol fuchsin (or 2% acetic orcein) to the watch glass.

(11) Place a coverslip on the slide and cover the slide with filter paper.

(12) Push down the coverslip with your thumb and the filter paper (This will squash the testes to burst the nuclei).

(13) Place the slide under a microscope and observe the stages of meiosis.

(14) Clean work area with surface disinfectant. Remove goggles and gloves and wash hands.

Ⅴ. Experimental Results

Get a prepared slide of the testes and ovary, and identify the meiotic stages you see. Make sketches of your observations.

Item	Visual field 1	Visual field 2	Visual field 3	Total	Percentage
Interphase					
Prophase					
Metaphase					
Anaphase					
Telophase					

Questions:

(1) List three major differences between the events of mitosis and meiosis.

In the mitosis, the nucleus is divided only once, and in the meiosis the nucleus is divided twice. Another difference is that mitosis produces two identical daughter cells, and meiosis produces four different daughter cells. Also, synapsis and crossing over do not take place in mitosis, but take place in the meiosis.

(2) Why is meiosis so important to sexual reproduction?

In the meiosis the chromosomal number is reduced to n so that it can be fertilized. Also, meiosis allows for chromosome crossing over, which can lead to the variations in organisms.

实验三　永久片的制作

一、实验目的

学习制作根尖永久片的基本方法。

二、实验原理

标本的观察通常是在显微镜下观察玻片标本。通常的步骤如下：首先将需要显微镜检查的样本放到载玻片上，然后着色，盖上盖玻片，这样样本就可以用于镜检了。镜检后如果玻片上的标本材料染色清晰，符合要求，就可用石蜡将盖玻片密封好，这样可以在冰箱里储存一个星期。如果希望保留更长一段时间，那就需要将玻片样本制成永久片。永久片的制作程序如下：先从载玻片上移除盖玻片，脱水，使之透明，用封固剂封固，然后盖上盖玻片。脱水技术采用一系列的梯度的醇、二甲苯以及二甲苯和封固剂混合溶液，最后用稀释的树脂封固。

三、实验材料

染色清晰符合要求的临时玻片标本、复式显微镜、蝗虫载玻片样片、拨针、载玻片、盖玻片、镊子、滤纸、刀片、培养皿、乙醇、50% 乙酸、异丙醇、正丁醇、二甲苯以及香树脂。

四、实验步骤

（一）移去盖玻片

（1）制作临时玻片标本（见上一实验）。

（2）玻片盖片用石蜡密封后，可置于冰箱中一周。

（3）从冰箱中拿出临时标本，用刀片擦掉盖玻片周围的石蜡，然后用浸过二甲苯的卫生纸擦拭剩余的石蜡。

（4）把玻片标本盖片朝下，倾斜着放入浸有液体的培养皿中（含等体积的95% 酒精和 45% 乙酸），让盖玻片自然脱落。

（二）脱水与透明

（1）按照下面的方法，将载玻片和盖玻片分别处理 30 s，风干。

①浸入 50% 乙醇中 30 s 后，风干。

②浸入 70% 乙醇中 30 s 后，风干。

③浸入 95% 乙醇中 30 s 后，风干。

④浸入 100% 乙醇中 30 s 后，风干。

（2）或按照下面的方法，将载玻片和盖玻片分别处理 5~10 min，风干。

①将盖玻片和载玻片迅速投入 1/2 冰醋酸 + 1/2 正丁醇中 5~10 min，风干。

②浸入正丁醇中 5~10 min 后，风干。

③浸入二甲苯中 5~10 min 后，风干。

（三）封片

（1）准备溶液（用 1/5~1/4 体积的二甲苯稀释石蜡）。

（2）拿出载玻片，用滤纸进行干燥（注意：不要碰到载玻片上的材料）。

（3）将一些上述准备溶液滴在材料上，盖上盖玻片。

五、实验结果

（1）每人制作清洁完整的永久片 1 张。

（2）从永久片制作成败的经验和教训中学到了什么？

注意事项：

当准备用来观察的玻片标本时，必须将所有需要的材料放在旁边，包括载玻片、盖玻片、滴管、移液管，以及任何计划使用到的化学药品或染色剂。

Experiment 3　Preparation of Permanent Microscopic Slides

Ⅰ. Experimental Objectives

To provide basic instructions for properly preparing permanent microscopic slides of root tips.

Ⅱ. Experimental Principles

A specimen is examined usually by using a slide for a microscope. The steps are as follows: Put the specimen, which will be examined, on the slide to stain. A cover is then put on the slide and now the specimen is ready for examination. The coverslip on the slide can be sealed with paraffin and stored in the refrigerator for a week if the material stained is clear and meets the requirements. The slide needs to be made as a permanent specimen if you wish to preserve it for a longer time. The permanent slide is prepared according to the following steps: removing the coverslip, dehydrating and making it transparent, coating with a mounting medium, and then putting a cover on the slide. The dehydrating techniques employ a series of gradient alcohols, xylene, and mixed xylene/mountant solutions. At the end the slides are sealed by the diluted resin.

Ⅲ. Experimental Materials

Temporary slides which are stained clearly and meet the requirements of the experiment. Compound microscope, prepared slides of grasshopper, dissecting needles, slides, cover slips, forceps, filter paper, scalpels, petri dish, alcohol, 50% acetic acid, isopropyl alcohol, butyl alcohol, xylene, and balsam.

IV . Experimental Procedures

Remove coverslips from slides

(1) Make the temporary slide (see the above experiment).

(2) The coverslip on the slide is sealed with paraffin, and it can be stored in the refrigerator for a week.

(3) Take the temporary slide from the refrigerator and polish the paraffin around the coverslip with a blade. Then wipe the residual paraffin by using tissue paper with dipping xylene.

(4) Then the temporary slide is tilted in a Petri dish (Containing the same volume of 95% ethanol and 45% acetic acid) with coverslip down ward, tilted, and let coverslip drop off naturally.

Dehydration and transparency

(1) Process the slide and coverslip separately through the following solutions for 30 seconds in each and dry:

① Immerse them into 50% ethyl alcohol for 30 seconds and dry.

② Immerse them into 70% ethyl alcohol for 30 seconds and dry.

③ Immerse them into 95% ethyl alcohol for 30 seconds and dry.

④ Immerse them into absolute ethyl alcohol for 30 seconds and dry.

Or

(2) Process the slide and coverslip separately through the following solutions for 5−10 min in each and dry:

① Immerse them into the solution (1/2 acetic acid + 1/2 butyl alcohol) for 5−10 min and dry.

② Immerse them into butyl alcohol for 5−10 min and dry.

③ Immerse them into xylene for 5−10 min and dry.

Sealing the slip

(1) Prepare the SOLUTION (dilute the paraffin by adding 1/5−1/4 volume of

Xylene).

(2) Take out the slide and dry by using the absorbent paper (**Attention**: Do not touch the slide materials).

(3) Drop the above prepared SOLUTION to the slide and cover the coverslip.

V . Experimental Results

(1) Every researcher should make one permanent slide.

(2) What can you learn from your success or failure of making permanent slide?

Attention:

When preparing microscope slides for observation, it is important to have all necessary materials at hand. These materials include slides, cover slips, droppers, pipettes and all chemicals or stains you plan to use.

实验四　果蝇形态及生活史的观察

一、实验目的

本实验的目的是向学生介绍实验室用于饲养果蝇的仪器和技术。在这个实验中，学生要了解果蝇的生活史，学会区分雄果蝇和雌果蝇的方法，以及怎样识别异常表型。

二、实验原理

果蝇为完全变态昆虫，其生活史包括卵、幼虫、蛹、成虫 4 个时期。果蝇和许多变温动物一样，其生长发育过程中所需的温度是不同的。理想的环境下果蝇在 25℃时发育时间是 8.5 天，在 18℃时将需要 19 天。雌果蝇能产 400 多个卵，每次产 5 个。卵产在腐烂的水果或其他合适的材料上，比如正在腐烂的蘑菇或有营养的液体。果蝇卵大小约 0.5 mm 长，25℃时卵需要 12~15 h 孵化。在 25℃下，孵化后幼虫生长约 4 天，在这期间，即卵孵化后在 24 小时和 48 小时分别蜕 1 次皮，而且它们以分解水果产生的微生物和水果本身的糖分为生，此后幼虫化成蛹。蛹经历长达 4 天蜕变（25℃下）即变成成虫。果蝇由于形态小，易获得易饲养，是实验常用材料。快速的繁殖周期使果蝇成为研究蜕变的适合物种。

三、实验材料

果蝇、过熟的香蕉、广口瓶（1L）、纸巾、橡皮筋、冰袋、培养皿、光学显微镜冰箱和孵化器。

培养基是一种复杂的混合物，包括琼脂、糖、其他营养补充剂，以及霉菌抑制剂。

四、实验步骤

（1）将香蕉剥开，放入广口瓶内，将广口瓶置于室外。

（2）观察广口瓶，看见一些果蝇飞进广口瓶后 30 min，盖住广口瓶，以防果蝇飞走。

（3）将广口瓶拿进室内，用纸巾覆盖瓶口，用橡皮筋扎住。

（4）将广口瓶置于室温下培养 14 天，其间不要动它。

（5）用透镜观察广口瓶里面的果蝇，记录其所有的变化。

（6）再将广口瓶放置两天让果蝇成熟，然后把广口瓶放入冰箱一小时，把果蝇杀死。将广口瓶拿出，摇晃广口瓶，把所有的果蝇都倒在一张纸上。

（7）用立体显微镜观察果蝇，记下雌性和雄性的个数。做两次重复实验，算出平均数。

（8）将广口瓶中的果蝇保留下来以做进一步研究。

（9）为了达到实验目的，要把果蝇放在塑料培养瓶中培养，塑料瓶里面最好有一小片塑料网用来增加表面积以使蛹附着在它的上面。

五、实验结果

（1）在这个实验中，你将观察到果蝇的整个生活史，确定变态各阶段的时间尺度以及温度对各变态阶段的影响。你需要检测出果蝇生长的理想环境并统计出每一代产生的雌雄果蝇数。

（2）记录表型：准确地验证形态特征，如眼睛的形状和颜色、体色、刚毛以及翅膀的形态。

（3）成年果蝇性别的确定：在记录的所有果蝇表型中，准确地确定每只果蝇的性别是极其重要的。在伴性遗传研究中，果蝇在遗传学中必须被分为雄性或雌性。

（4）果蝇生活史的观察：果蝇为完全变态昆虫，其生活史包括卵、幼虫、蛹、成虫 4 个时期。

问题：

（1）除了完成这个实验以外，如何区分雄蝇和雌果蝇？

雌性和雄性的果蝇能被鉴别出来：雄果蝇体形明显小，末端钝，雄性果蝇的腹部后端有黑斑，一直延伸到前侧的两边。雌果蝇有细的黑色的线条穿过腹部，但仅限于腹部上面，下面没有。注意：性别很难根据颜色区分，除非是几个小时大的果蝇，因为刚孵化的果蝇很暗淡。

（2）每一代果蝇中，雌比例是多少？每一代果蝇中，雄比例是多少？

（答案略）

附录1：

果蝇的生活史

在香蕉表面上，从看起来湿漉漉的小白色斑点变成白色幼虫需要两天。四天或五天以后，可以观察到像麦粒颜色和形状一样的东西附着在广口瓶内壁。十天后，果蝇就会出现。果蝇的蜕变包括四个阶段，在第一个阶段——卵，为椭圆形，其中一个末端有两个突起。这两个突起能使它漂浮在液体培养基上。第二个阶段发生在第一阶段后的 24~48 h，此时卵变成了幼虫。幼虫呈白色，常被称为蛆。随着不断摄食，幼虫能在水果中爬行，约 4 天以后，幼虫爬到广口瓶的边缘，缩小，形状变得更椭圆。这时，进入第三个阶段或者说蛹的阶段。此时仍是白色的，但几个小时后，颜色会变深。在蛹阶段，看不到变化，但是在这个阶段，足、翅和头长成。约十天以后，到达第四个，也就是最后一个阶段，即颜色变为乳白色的果蝇诞生了。果蝇的颜色很快就会变暗。

卵：卵长约 0.5 mm，卵形，白色。从背侧前表面延伸有一对卵丝，卵丝的腹面稍扁平，像漂浮的手，可以避免其沉入半液体培养基中。

幼虫：幼虫从卵里孵化出来，呈白色的、扁形和蠕虫状。头窄，嘴尖部分呈黑色。幼虫后要经两次蜕皮，才能从一龄幼虫发育到三龄幼虫。在这个阶段之后，幼虫爬出培养基，最终贴附在瓶壁上形成蛹。

蛹：不久在蛹的前部呼吸孔形成，然后幼虫的身体缩短、皮肤变硬和色素出现。在这个过程中，大部分的成体结构形成。一个完全的成蝇通过蛹的前端出来。此羽化时期，成蝇较长，颜色光亮，翅还未展开。此后翅膀很快展开，身体逐渐变成黑色或棕色。从蛹中出来 6 h 后，成虫便有了参与繁殖的能力。

成虫：身体分为头部、胸部和腹部。头部有一对复眼，一对触角。胸腔分为 3 段：前胸、中胸和后胸。每一段都有对腿。有一对翅在中胸。雄性腹部有 4~5 节，雌性有 6~7 节。雄性的腹尖为黑色。

雄性和雌性果蝇之间的区别

项　目	性　状	雄　性	雌　性
1	身体大小	小	大
2	腹侧	3 节	5 节
3	腹尖	有	无
4	腹尖性状	圆	尖
5	前腿性梳	有	无

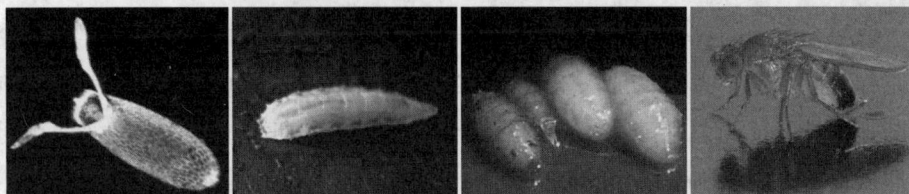

卵　　　　　　　幼虫　　　　　　　蛹　　　　　　　成虫

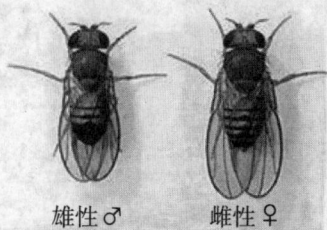

雄性♂　　　雌性♀

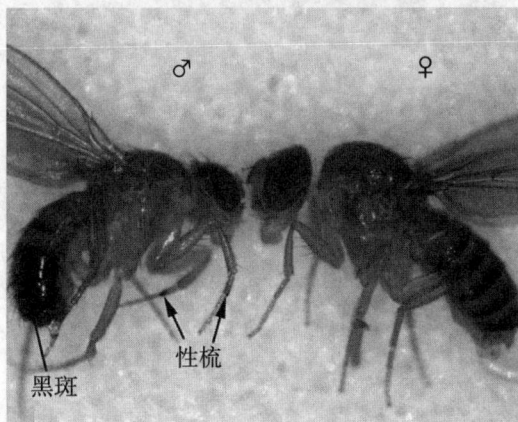

♂　　　　　　　　♀

黑斑　　　性梳

附录 2：

果蝇培养基配方如下：

成　分	数　量	单　位	作　用
蒸馏水	80	mL	溶剂
玉米粉	8.2	g	基本成分
白糖	6.2	g	基本成分
琼脂条	0.62	g	固化剂
丙　酸	0.5	mL	防腐剂
酵母粉	少量		菌种

配法：

取一个干净的容器，加水 100 mL 烧开。然后加入 6 g 的糖，拌匀。加 0.62 g 琼脂作为固化剂，煮沸，加入玉米粉 8.2 g。然后添加丙酸 0.5 mL，作为抗菌剂。不断地搅拌，成为黏性流体，立即将混合物转入培养瓶。该瓶冷却时加入酵母，并插入棉塞。用于制备培养的瓶应消毒，棉塞同样也要消毒。随着培养基养分条件变坏，至少每 3 周将果蝇转入含有新培养基的瓶子中。

Experiment 4   Observation on Drosophila and Its Life Cycle

Ⅰ. Experimental Objectives

The purpose of this experiment is to introduce to students the equipment and techniques of feeding "fruit flies" in the laboratory. You will learn about the life cycle of the Drosophila, also learn to differentiate male and female flies and learn how to identify aberrant phenotypes.

Ⅱ. Experimental Principles

Drosophila is the complete metamorphosis insect and its life history includes the four periods: eggs, larvae, pupae, and adults. The developmental period for *Drosophila melanogaster* varies with temperature, like the other ectothermic species. Under ideal conditions, such as at 25℃ , the development time of *Drosophila melanogaster* is 8.5 days, and at 18℃ , it takes 19 days. Females lay about 400 eggs (embryos), around five at a time, in rotting fruit, mushrooms, sap fluxes or other suitable materials. The eggs, which are about 0.5 millimeters long, need 12−15 hours to hatch (at 25 ℃). The larvae grow continually about 4 days (at 25℃). During this period, they molt twice (into 2nd- and 3rd-instar larvae) at about the 24th and the 48th hour respectively and live on some microorganisms in decomposing fruit, as well as on the sugar of the fruit itself. Then the larvae encapsulate in the puparium and undergo a four-day metamorphosis (at 25℃). After that the adults enclose (emerge). Fruit flies (*Drosophila melanogaster*) are often used as research specimens because they are small, and easy to get and maintain. The fast reproductive cycle makes them be the most favorable specimens for studying the process of metamorphosis.

III . Experimental Materials

Drosophila, overripe banana, (1-liter) jar, paper towel, rubber band, ice packs, petri dishes, a light microscope, a refrigerator, and an incubator.

The culture medium is a complex mixture of agar, sugars, other nutritional supplements, and mold inhibitors.

IV . Experimental Procedures

(1) Peel the banana and place it into the open jar. Set the jar outside.

(2) Observe the jar. Wait 30 minutes and cover the jar to prevent the flies flying away after several flies are seen in the jar.

(3) Bring the jar in the room. Cover the mouth of the jar with the paper towel and secure the paper with the rubber band.

(4) Allow the jar to stand undisturbed at the room temperature for 14 days.

(5) Use the magnifying lens to observe the contents in the jar. Make notes of all changes.

(6) Wait another two days to allow the flies to mature and then put the jar into a freezer for one hour to kill the flies. Take out the jar and shake the flies out of the jar on a sheet of paper.

(7) Use a stereo microscope to observe the flies and count the number of each sex. Repeat two times of the experiment and calculate the average percentage of each sex.

(8) Keep the flies in the jar for further research.

(9) For our purpose, D. melanogaster is raised in plastic culture vials which may contain a small piece of plastic netting to increase surface areas for the attachment of pupae and also the culture medium.

V . Experimental Results

(1) In this project, you will observe the general life cycle of the fruit fly. The timescale of each stage of the metamorphosis and effects of temperature on each stage will be determined. You will find the desired environment for the development of the fruit

flies as well as the percentage of the males and females produced in each generation.

(2) Scoring the phenotypes: It will often be necessary to accurately verify the phenotypic characteristics such as eye color and shape, body color, and bristle as well as wing morphology.

(3) Determining the sex of adult flies: When scoring various phenotypes, it is critical that the sex of each fly be accurately determined. Flies must be classified as males or females in sex linkage studies.

(4) Study on the life history of Drosophila: It is the complete metamorphosis insect, amd its life history includes the four periods of eggs, larvae, pupae and adults.

Questions:

(1) Upon completion of this exercise, how can one distinguish male and female *D. melanogaster*?

Male and female flies can be identified. The male fly is noticeably smaller and has a pointed posterior. The rear of the male's abdomen has intense black coloring that extends around the sides to meet ventrally (in the front, near the bottom). The female has small, black-colored lines across the abdomen that are confined to the upper part and never appear underneath. Note: Sexes are difficult to distinguish from their colors until the flies are a few hours old for newly hatched flies are very pale.

(2) What is the percentage of the females in each fly generation? What is the percentage of the males in each fly generation?

(The answer is omitted.)

Appendix 1:

Drosophila life cycle

Small, white, moist-looking specks on the surface of the banana change into white wormlike creatures within two days. At the end of four or five days, objects of the color and shape of wheat grain can be seen stuck to the inside wall of the jar at various heights. Flies appear in the jar in another ten days. The metamorphosis of the fly involves four separate stages. In the first stage, there is the egg, which is elliptical and has two small projections on one end. These projections enable it to float in liquid mediums. In the second stage, which occurs within 24 to 48 hours after the first stage, the egg develops into a larva. The larva is the white wormlike organism commonly called a maggot. By eating continually, the larva makes channels in the fruit. About four days later, the larva crawls to the side of the jar, contracts, and becomes more elliptical in shape. At this time, it enters the third, or pupa, stage. The larva is still white; however, within hours it darkens. No motion is observed during the pupa stage, but it is during this period that legs, wings, and a head take shape. About ten days later, the fourth and final stage is reached and a pale-colored, fragile fly emerges. Shortly thereafter, the flies darken in color.

Egg

Egg is about 0.5 mm in length, ovoid in shape, and white in color. Extending from the anterior dorsal surface, there is a pair of egg filaments. The terminal portion of these filaments is flattened into spoon like floats. These floats keep the egg from sinking into the semi-liquid medium.

Larva

The larva hatches out from the egg. It is white, segmented, and wormlike. The head is narrow and has black mouth parts. The larva undergoes 2 molts, so that the larva develops from the 1st instar to the 3rd instar. After this stage, the larva crawls out of the medium and finally attaches to the inner drier surface of the bottle. This culminates in pupation.

Pupa

Soon after the formation of the "pupal horn" from the anterior spiracle, the larval body is shortened and the skin becomes hardened and pigmented. During this process, most of the adult structures are developed. A fully transformed adult fly emerges out through the anterior end of the pupal case. At the time of eclosion, the fly is greatly elongated and light in color, with wings yet unfolding. Immediately after this, the wings unfold and the body gradually turns dark or brown. After 6 hours of emergence, the adult fly attains the ability to participate in reproduction.

Adult

The body consists of head, thorax, and abdomen. The head has a pair of compound eyes and a pair of antennae. The thorax is divided into 3 segments: prothorax, mesothorax, and metathorax, each with a pair of legs. The mesothorax has a pair of wings. The abdomen is segmented into 4 or 5 sections in males and 6 or 7 in females. The abdominal tip of males is darkly pigmented.

Differentiation Between Male and Female Drosophilas

Item	Character	Male	Female
1	Body size	Small	Large
2	Dorsal side of abdomen	3 separate	5 separate
3	Abdominal tip	Present	Absent
4	Abdominal tip shape	Round	Pointed
5	Sex comb in foreleg	Present	Absent

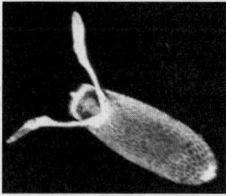  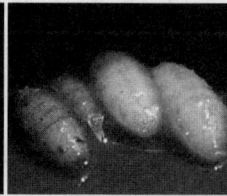 

Egg   Larva   Pupa   Adult

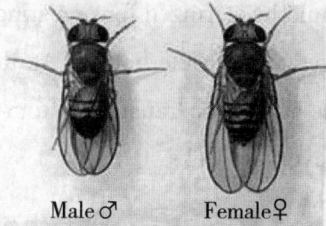

Male ♂   Female ♀

Sex   comb

Blackspot

Appendix 2:

The gradients of the culture medium and how you make the culture medium are as follows.

Component	Weight	Unit	Function
Distilled water	80	mL	Solvent
Corn flour	8.2	g	Basic ingredient
Sugar	6.2	g	Basic ingredient
Agar	0.62	g	Solidifying agent
Propionic acid	0.5	mL	Preservative
Yeast granule	a little		Strain

How do you prepare the culture medium?

Take a clean vessel and boil 100 mL of water, then add 6 g of sugar and stir well, at last add 0.62 g of agar, which acts as a solidifying agent. Once it boils, add 8.2 g of corn flour, and then pour 0.5 mL of propionic acid, which acts as an antimicrobial agent. By constant stirring, the medium becomes a viscous fluid. The hot mixture is transferred into the culture bottle which is left for cooling. Add yeast into the bottle, and plug the bottle with cotton. Bottles should be sterilized for preparing culture medium. Similarly, sterilized cotton has to be used to plug the bottles. As the condition of the medium deteriorates with time, the flies have to be transferred from old bottles to new ones with fresh culture medium at least every 3 weeks.

实验五　果蝇唾腺染色体的观察

一、实验目的

本实验有三个目标：第一，学习从果蝇幼虫中分离唾液腺。第二，制作多线带染色体涂片。第三，观察在染色体加热后膨大的多线染色体。

二、实验原理

在果蝇幼虫早期，唾腺细胞构建后，其细胞分裂就停止了。在细胞大小增长的同时，细胞核也在增长，其染色体大量复制而不进行细胞分裂。在细胞分裂周期的间期，其染色体伸展并达到最大长度。此时多线染色体有 1 000~4 000 染色体拷贝，比其他果蝇细胞染色体长 100~200 倍。多线染色体有如此多的 DNA 拷贝，因此它们在光学显微镜下容易被看见。这些染色体的重要特征是其有明暗相间的条纹，像条形码一样，并且每一条染色体都是独一无二的。暗条纹出现的地方 DNA 比较密集，而亮条纹（间带）出现的地方 DNA 较为稀疏。这些条纹显示的可见标记可以被用来识别染色体特定基因的位置或染色体上的位点。

三、实验材料

双目立体显微镜、复式显微镜、拨针、手术刀、载玻片、盖玻片、滤纸、大龄幼虫、1%醋酸地衣红、卡宝品红、45% 醋酸、1 mol/L HCl、蒸馏水以及乳酸—45% 乙酸溶液（1∶1）。

四、实验步骤

（一）果蝇唾液腺的挑取步骤

（1）从储藏的果蝇中取大个的果蝇幼虫，大个的果蝇幼虫容易解剖，也可选取没有化蛹的活体幼虫。

（2）在双筒解剖镜下，用解剖针在其口的后端和前端部分用针尖固定住幼虫，然后解剖幼虫。

（3）有两个透明唾液腺位于幼虫的前端。腺体的形状像颗粒状物，将围绕腺体的狭窄的、白色条带状的脂肪去掉。

（4）留下唾液腺，其余不要。

（二）染色和观察

（1）将唾液腺置于载玻片上，滴 1 滴 45% 乙酸溶液于腺体上，5~7 s 后用滤纸小心地吸净剩余溶液。

（2）再滴 2 滴 1 mol/L HCl 于腺体上，30~40 s 后用滤纸小心地吸净剩余溶液。

（3）滴 1 滴蒸馏水于腺体上，1~2 s 后用滤纸吸干。

（4）滴 2 滴卡宝品红（或乳酸—乙酸—地衣红）于腺体上，染色 10~20 min。

（5）滴 2 滴乳酸—乙酸（45%）3 次，用滤纸吸干。

（6）盖上盖玻片，用拇指垫上吸水纸压碎腺体，破裂核膜，释放染色体。先在低倍镜下观察，找到云雾状的染色体区域后，换到高倍镜下观察。

五、实验结果

（1）每个学生都需要准备自己的果蝇唾腺玻片。

（2）绘出自己所观察到的果蝇唾腺染色体的图像。

注意事项：

这些实验能否成功在很大程度上取决于发育阶段的幼虫。如果它们太小，那么它们的唾液腺染色体还没有发育到足够好。随着幼虫不断变大到蛹化，唾液腺开始退化，不再适合用作染色体的样本。尽可能选择大的、有活力的幼虫。不要选择比其他幼虫颜色明显较深的幼虫，深色的表皮是幼虫蛹化的迹象。

Experiment 5　　Observation on Drosophila Polytene Chromosomes

Ⅰ. Experimental Objectives

There are three objectives for this experiment: First, to isolate the salivary glands from Drosophila larvae. Second, to prepare stained smears ("squashes") of banded polytene chromosomes. Third, to observe polytene chromosomes puffing after the chromosomes have been heated.

Ⅱ. Experimental Principles

After Drosophila salivary gland cells are established during early larval development, cell division ceases. When the cells increase in size, their nuclei grow too, as the chromosomes duplicate repeatedly without accompanying cell division. The chromosomes are in an extended interphase of the cell cycle and, as such, are stretched out to their full length. The polytene chromosomes are 100-200 times longer than the other chromosomes of Drosophila cell and reach 1000-4000 copies of chromosome. Because polytene chromosomes have so much DNA, they are easily visible under the light microscope. A useful feature of these chromosomes is that they have a pattern of dark and light bands, like a bar code, which is unique to each chromosome. The dark bands represent regions where the DNA is most densely packed, and the light bands (interbands) are regions where the DNA is less densely packed. These bands provide visible landmarks that can be used to identify the location of a specific gene on the chromosome or the sites of chromosomes.

III . Experimental Materials

Binocular stereomicroscope, compound microscope, dissecting needles, scalpel, microscope slides, cover glasses, filter paper, Drosophila virilis larvae, 1% aceto-orcein stain, carbol fuchsin, 45% acetic acid solution, 1 mol/L HCl, distilled water, and lactic acid-45%-acetic-acid solution (1 : 1).

IV . Experimental Procedures

Procedure for removing Drosophila salivary glands

(1) Remove a large larva from the stock of *D. virilis*. Larger larvae are easier to dissect. Selecting an active larva which has not started to pupate is also adoptable.

(2) Using the stereomicroscope, dissect the larva by placing one dissecting needle on the posterior aspect of the larva and the other needle at the anterior end, near the black mouth parts.

(3) There are two transparent salivary glands located anteriorly in the larva. The glands are characterized as a granular, bead-like appearance. A narrow, white ribbon of fat which surrounds the glands should be torn away.

(4) Discard the entire larva except for the salivary glands.

Staining and observing

(1) Place the salivary glands on the slide and drip one drop of 45% acetic acid solution on the salivary glands. After 5−7 seconds, dry them with a paper towel.

(2) Drip 2 drops of 1 mol/L HCl on the salivary glands, and let it stand for 30−40 seconds. Dry them with a paper towel.

(3) Drip one drop of distilled water on the salivary glands. After 1−2 seconds, dry them with a paper towel.

(4) Drip 2 drops of carbol fuchsin (aceto-orcein) on the salivary glands, and let it stand for 10−20 minutes.

(5) Drip 2 drops of 45% acetic acid solution on the salivary glands for three times and dry them using a paper towel.

(6) Place a cover slip over the glands, and by using your thumb and a paper towel push down on the slide. The pressure applied will squash the glands, rupture the nuclear membrane, and free the chromosomes. Using a compound microscope, observe the slide under low and high magnification.

V . Experimental Results

(1) Each student should prepare his or her own slides of polytene chromosomes.

(2) Draw the polytene chromosomes as you see them under the microscope.

Attention:

The success of these experiments is determined to a large extent by the developmental stage of the larvae. If they are too small, their salivary glands aren't sufficiently developed to yield good chromosomes. As the larvae become larger and approach pupation, the salivary glands begin to degenerate and are no longer suitable for chromosome samples. Choose the largest larva that you can find and that is still moving actively; do not choose a larva that is noticeably darker than the rest, as darkening of the cuticle is a sign that pupation is imminent.

实验六　果蝇的伴性遗传

一、实验目的

在本实验中我们通过分析红眼果蝇与白眼果蝇杂交来理解伴性遗传的模式。

二、实验原理

一般情况下，果蝇个体的性别是由一对性染色体决定的。雌性的性染色体是纯合子，雄性则是杂合子。伴性遗传定义指由性染色体参与的个体性状特征的遗传，而这种性状特征就被描述为伴性特征。伴性现象首先由摩尔根在 1910 年对果蝇进行研究时发现。他发现正常野生型眼睛的雄性果蝇中出现白色眼，因此他提出了伴性遗传这个现象。

果蝇有四对染色体，第一对为性染色体，其余三对为常染色体。果蝇的性别决定是 XY 为雄，XX 为雌。果蝇中已知红色眼和白色眼是一对相对性状，由位于 X 染色体上的基因（ + /W）决定的，而 Y 染色体没有对应的等位基因。

三、实验材料

果蝇（*Drosophila melanogaster*）红眼和白眼品系，包含培养基的小瓶、冰箱、冰盒、培养皿、显微镜、放大镜、果蝇盛留瓶、饲养瓶、解剖针和笔、纸。

四、实验步骤

红眼果蝇和白眼果蝇分别培养。当蛹在瓶中出现时，将瓶里的果蝇都取出。从蛹开始，处女蝇分开饲养 2~3 天，然后按如下方法杂交：

红眼雌果蝇 × 白眼雄果蝇。

红眼雄果蝇 × 白眼雌果蝇。

观察统计 F_1 的表型并记录数据。一些 F_1 近交产生 F_2 代，观察统计 F_2 的表型并记录数据。

五、实验结果

收集至少 200 个果蝇，记录下果蝇的性别和特征，并用卡方检验分析数据（详见卡方分析实验）。

表　型	观察值	期望值	Deviation	d^2	$d^2/E = \chi$
红眼 (male)					
红眼 (female)					
白眼 (male)					
白眼 (female)					
$\Sigma \chi^2 =$					
自由度 Degree of freedom =　 $-1 =$					

问题

（1）描述观察到的突变果蝇眼的颜色。

在 F_1 代中雄性果蝇的眼睛是白色的，雌性果蝇的眼睛是红色的。在 F_2 代中，雄性和雌性的眼睛可能都是红色或者白色的。

（2）写下你研究的性状遗传模式的假设。

先假设你的遗传模式为无效假设。伴性遗传在表型上总是以 1∶1 的比例。在 F_1 代中，红眼雌果蝇与白眼雄果蝇的比例是 1∶1。在 F_2 代中，红眼雌果蝇与白眼雌果蝇的比例将会是 1∶1；其比例在所有雄果蝇中也是这样的，从下表中可以计算出该结果。

项　　目	X^+	Y
X^+	X^+X^+ 红眼 ♀	X^+Y 红眼 ♂
X^W	X^+X^W 红眼 ♀	X^WY 白眼 ♂

Experiment 6　Sex-linked Inheritance of Drosophila Melanogaster

Ⅰ. Experimental Objectives

In the present experiment, we have taken the red eye mutant and crossed it with a white eye strain to study the pattern of sex-linked inheritance.

Ⅱ. Experimental Principles

In the majority of cases, the sex of an individual is determined by a pair of genes on the sex chromosome. Females are homozygous and males are heterozygous. Sex-linked inheritance is defined as the inheritance of somatic characters, which are linked with sex chromosomes. The characters are described as sex-linked characters. The phenomenon of sex linkage was first observed by T. H. Morgan in 1910 who experimented on Drosophila. Morgan observed the appearance of white eye color in males when he cultured the normal wild-eyed flies. He thus proposed the phenomenon of sex linkage.

Fruit flies have four pairs of chromosomes, of which the first pair is sex chromosomes, and the remaining three are autosomal. Sex determination in Drosophila is XY and XX for male and female respectively. The red eye and white eye of Drosophila is a pair of relative characters which is controlled by the genes located on the X chromosome (+ /W) while Y chromosome has no corresponding allele.

Ⅲ. Experimental Materials

The materials used in this experiment are as follows: Drosophila with red eye and white eye trait, vials containing a medium, a refrigerator, ice packs, Petri dishes, a light microscope, a vial of wild type flies, an incubator, dissecting needles, a pencil and some paper.

IV . Experimental Procedures

Drosophila melanogaster: The red-eyed and the white-eyed flies are cultured in standard media separately. When the pupa appears in the bottles, the bottles are cleaned by taking out all the flies presented. From the beginning of pupa, the virgin females are isolated until aged 2−3 days, and then the crosses are conducted as follows:

Red-eyed females × white-eyed males.

Red-eyed males × white-eyed females.

The phenotype of the progeny produced in the F_1 is observed and the data is recorded. Some F_1 progenies are inbred to yield the F_2 generation, of which phenotype is also observed, and the data is recorded.

V . Experimental Results

Try to collect at least 200 flies. Record their sex and characteristics.In order to analyze your data you will first have to complete the table of Chi-Square analysis (See the experiment of Chi-Square analysis).

Phenotype	Observed	Expected	Deviation	d^2	$d^2/E = \chi$
Red-eyed (male)					
Red-eyed (female)					
White-eyed (male)					
White-eyed (female)					
$\Sigma \chi^2 =$					
Degree of freedom = −1=					

Questions:

(1) Describe the eye color which you observe from the mutations.

In the F_1 generation the males have white eyes and the females have red eyes. In the F_2 generation, the males and females could have either red or white eyes.

(2) Write a hypothesis which describes the mode of inheritance of the trait you studied.

This is your null hypothesis. For a sex linked phenotypes are always a one to one ratio. In the F_1 generation, the ratio of the number of red-eyed females to white-eyed males is one to one. In the F_2 generation, the ratio of red-eyed females to white-eyed females is one to one and their ratio in males is also the same. You can reach the conclusion from the following table.

Item	X^+	Y
X^+	X^+X^+ red-eyed ♀	X^+Y red-eyed ♂
X^w	X^+X^w red-eyed ♀	X^wY white-eyed ♂

第二部分

实验七　人类染色体的识别与核型分析

一、实验目的

（1）学习染色体核型的分析方法。

（2）了解人类染色体的特征。

二、实验原理

尽管在 20 世纪末人们就观察到了染色体，但是直到 1902 年，西奥多·博韦里和沃尔特·萨顿才把细胞分裂中的染色体行为和遗传物质行为联系起来。染色体的细胞学技术发展，使得我们可以调查染色体数目、形态以及形态与遗传现象的关系。在细胞周期过程中，染色体在形态上总是在变化。在分裂中期染色体高度浓缩，在这个时候，每对染色体都具有其特定的形态特性。在标准光学显微镜下，我们可以观察到着丝粒的相对大小、位置以及微卫星等特性。因为不同的染色体往往有非常相似的形态特征，所以这种技术通常无法区分所有染色体核型，尤其形态较小的染色体。

为了便于识别各种不同的染色体，各种染色技术已经发展起来。早期观察的染色体都是异染色质，在异染色质上细胞间期着色区和非着色区便能呈现出来。这些染色模式似乎与染色体上蛋白质和 DNA 顺序精确布置，以及活跃和不活跃的基因变化有关。C 带是染色体着丝粒区（着丝粒异染色质）上的深色区域，它可以区分长度上相似的染色体。有证据表明，异染色质区域基因不活跃。有些荧光染料对 DNA 亲和，在紫光灯下能显示明暗条带。最广泛使用的荧光染料是喹吖因芥子，它所产生的条带模式被称为 Q 带。在学生实验中，Q—谱带不方便使用，因为它们需要用荧光显微镜才能检测，而且荧光衰减得很快。蛋白水解酶水解和 Giemsa 染料染色结合使用是目前最为广泛使用的技术，它能产生最具特色的 G 带。其反向过来便是 R 带，它是通过吖啶橙染色产生的，它显示出了 G 带的反向条带。利用这种方式，核型结果就可以用于研究染色体畸变，如缺失、重复、易位、倒置，或者不寻常的染色体。这些技术往往能用于疾病特征的诊断。

三、实验材料

人类染色体永久载玻片、复式显微镜、尺子，以及剪刀等。

四、实验步骤

（1）将玻片放置于显微镜的载物台上，并观察载物台上永久载玻片的染色体。

（2）选择完整的细胞，完整的细胞具有清晰的轮廓，其形态较好，染色体分布在同一水平面。

（3）用数码相机拍摄显微镜照片，并进行分析。

五、实验结果

（1）观察标准细胞 20~50 个。

（2）显微照片分析方法如下：

着丝点指数（%）=（短臂/整套单倍染色体总长）× 100%= $\dfrac{}{p\ q}$ ×100%

臂比率 $=q/p$

（3）配对：根据染色体大小、着丝点位置和随体有无配对。

（4）染色体排列分组：p 向上，q 向下，着丝点排列在一条直线上。

（5）制作核型分析板，作出书面报告。

从染色体照片来制作人类染色体核型，具体方法是通过眼睛鉴定染色体，用剪刀将其剪出，将其分成不同的小组，并粘贴好。

（6）根据所给定的染色体图片进行测量，将结果填写在下表里。

编号	绝对长度	相对长度	短臂	长臂	臂比率	着丝粒指数	随体	类型

附：

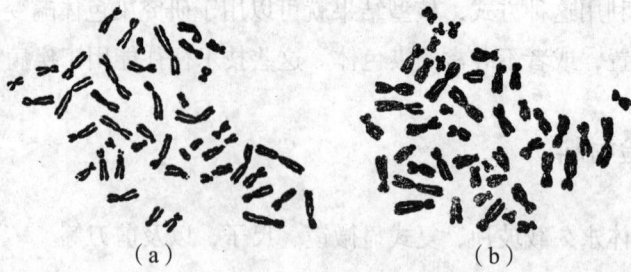

图1　人类染色体

（a）男性；（b）女性

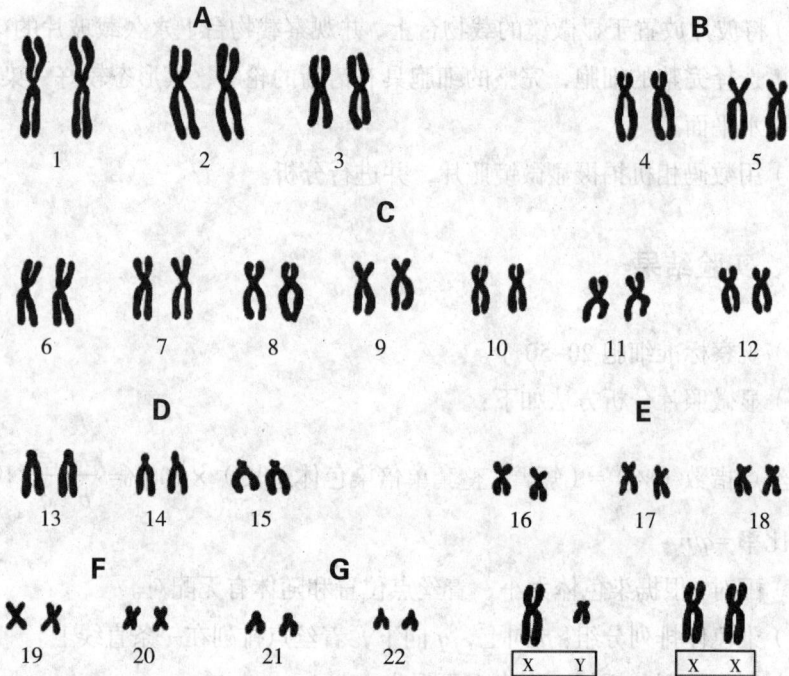

图2　人类标准核型（2n=46）

除了性染色体外，男性和女性核型相同，男性和女性的性染色体分别是XY，XX。基于大小和着丝粒位置差异将染色体分为7组，这7组如下：

A：大型中间着丝粒

B：大型近端着丝点

C：中型中间着丝粒

D：中型端着丝点

E：小型近端着丝点

F：小型中间着丝粒

G：小型端着丝点（Y染色体存在于此组）

Experiment 7   Identification and Karyotype Analysis of Human Chromosomes

I . Experimental Objectives

(1) Study the methods of analysis of karyotype

(2) Understand the characteristics of human chromosomes.

II . Experimental Principles

Although chromosomes were observed in the end of the last century, it was not until 1902 that Theodor Boveri and Walter Sutton associated their behavior in cell division with the behavior of the genetic materials. The development of cytological techniques for chromosome study has allowed investigations into chromosome numbers, morphology, and the relation of morphology to genetic phenomena. Chromosomes are variable in morphology over the course of the cell cycle, and are most highly condensed during metaphase. At that time, each member of a chromosome pair has a specific morphological character. Standard light microscope preparations can show such identifying characteristics as relative size, location of centromere, and the presence of satellites. Because different chromosomes often have very similar morphologies, such techniques are typically unable to distinguish all chromosomes in a karyotype, especially the smaller ones.

To facilitate the identification of individual chromosomes, a variety of staining techniques have been developed. An early observation is that there are regions (heterochromatin) on the chromosome that take up various dyes and others that remain unstained (euchromatin) during interphase. These patterns appear to be due to variation in the exact arrangement of protein and DNA along the length of the chromosome, and to the activity or inactivity of genes. C-bands detect darkly-staining material at the chromosomal centromeres (centromeric heterochromatin), which can differentiate chromosomes of similar length. Evidence indicates that heterochromatic regions are genetically inactive. Certain fluorescent dyes have an affinity for DNA and under ultraviolet light show light and dark bands. The most widely used of these is quinacrine mustard, and the banding patterns produced are known as Q-bands. Q-bands are inconvenient for student labs both because they require a fluorescence microscope, and because the fluorescence fades quickly. Perhaps the most widely used technique is a combination of proteolytic enzyme digestion and staining with Giemsa dye, which produces highly characteristic pattern known as G-bands. Reverse, or R-bands, can be produced by acridine orange, which shows a reverse of the banding pattern of G-bands. Once arranged in this way, the resulting karyotype may be used in the study for chromosomal aberrations such as deletion, duplication, translocation, inversion, or an unusual chromosome number. These often produce diagnostic suites of medical conditions (syndromes).

III. Experimental Materials

Prepare the permanent microscopic slide of human chromosome, compound microscope, ruler and scissors, etc.

IV. Experimental Procedures

(1) Place the slide on the stage of a microscope and view the stages of chromosomes.

(2) Choose the cell which is integrate and has a clear outline. The chromosomal morphology is relatively good and its distribution is in the same horizontal plane in this cell.

(3) Photograph by using the digital photographic microscope for analysis.

V. Experimental Results

(1) Observe 20−50 of the standard cells.

(2) Micrograph analysis:

Centromere index (%)——the percentage of the short arm in the whole chromosome.

$$\frac{p}{p+q}\times 100\%$$

Arm ratio $=q/p$

(3) Pair according to the size, centromere position and trabant of chromosomes.

(4) Chromosome alignment: p−upwards; q−downwards.

(5) Do the Karyotype analysis and write a report.

Human karyotypes are prepared from photographs of chromosome spreads, of which the individual chromosomes are identified by eyes. Cut out the chromosomes with scissors, physically sort them into groups, and stuck them into place with double-sticky tape.

(6) According to the data of the chromosome images given, fill in the table below.

Number	Absolute length	Relative length	Short arm	Long arm	Arm ratio	Centromeric index	Satellite	Type

Appendix:

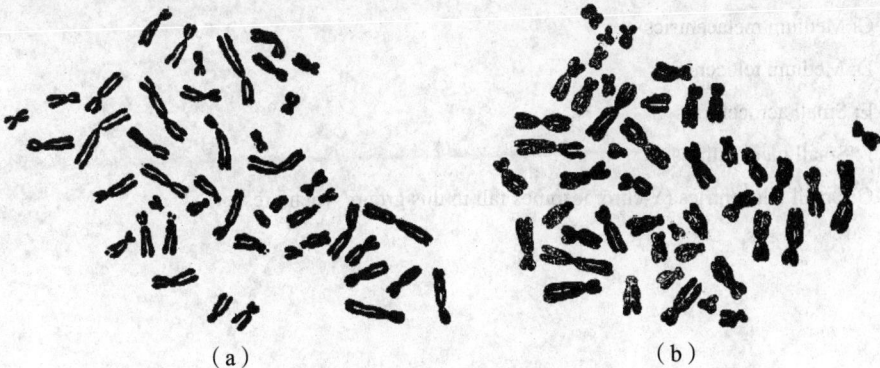

Figure 1   Human chromosomes

(a) For men; (b) For women

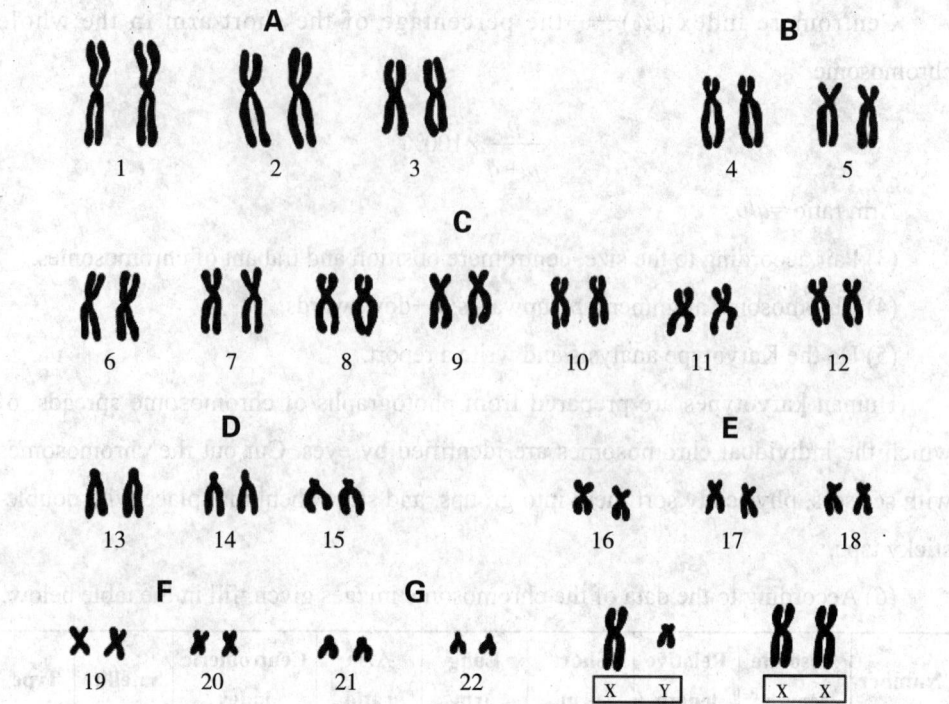

Figure 2   Standard karyotypes of Homo sapiens (2n=46)

Male and female karyotypes are identical, except the sex chromosomal pair which is XY in males and XX in females. Chromosome pairs are divided into seven groups based on the size and the centromere position: These are grouped as follows.

A: Large metacentrics

B: Large acrocentrics

C: Medium metacentrics

D: Medium telocentrics

E: Small acrocentrics

F: Small metacentrics

G: Small telocentrics (Y-chromosomes fall in this group, when present)

实验八　人体性染色质体的观察

一、实验目的

通过实验观察人类的性染色质，学习染色质制备和观察分析方法，进一步了解异染色质、X 染色体失活等相关知识。

二、实验原理

在哺乳动物中，雌性具有含两条 X 染色体上的所有基因位点，而雄性只有其中的一个染色体上的拷贝位点，因此人们认为性连锁的基因产物表达量上，雌性是雄性的两倍。然而一个称为剂量补偿的机制阻止了基因被这样表达。哺乳动物受精后约 16 天时，所有体细胞中的一条 X 染色体失活。雌性或雄型亲本的 X 染色体的失活是随机的。如果一个个体的性连锁基因位点是杂合子，这在不同的细胞中意味着其中一个位点将有选择地失活，而从它们衍生的后代细胞中将只表达一个或另一个等位基因，而不是都表达。这些个体在遗传上是嵌合体。从细胞学看，失活的 X 染色体浓缩成的核酸小体又称性染色质或巴小体（它的发现者之一为默里·L·巴尔，加拿大的细胞遗传学家）。典型的 XX（雌性）有一个巴小体，典型的 XY（雄性）没有巴氏小体。在超过一个 X 染色体数量的个体中，所有的染色体中的一个 X 染色体浓缩成巴小体。比如怀疑一些东欧国家企图把男性运动员变成女性，在奥运会上毛囊中的巴小体可用于性别测试。

三、实验材料

口腔黏膜、复式显微镜、拨针、载玻片、盖玻片、镊子、滤纸以及刀片。
乙醇、50% 乙酸、2% 醋酸地衣红、盐酸以及卡宝品红。

四、实验步骤

（1）用自来水漱口 3 次，除去口腔中的部分细菌和一些将脱落的上皮细胞。

（2）将玻片彻底洗干净：先用洗涤剂洗，然后用无菌水冲洗，再用 70% 乙醇冲洗，最后烘干。

（3）用清洁灭菌的牙签，从口腔两侧颊部刮取上皮黏膜细胞，分别涂抹在干净载片上，并且自然干燥。

（4）滴加 95% 的乙醇 2 min。滴加 70% 的乙醇 2 min。滴加无菌水 2 min。

（5）滴加 5 mol/L 或 6 mol/L HCl 大约 5 s。这一步将除去 RNA（时间很关键）。

（6）在水槽上用无菌水冲洗 10~15 s。

（7）滴加 2% 醋酸地衣红 / 卡宝品红，在室温下染色 10~15 min。

（8）无菌水冲洗。

（9）加盖玻片，在低倍镜下定位，然后在高倍镜或油镜下观察。

五、实验结果

观测至少 20 个男性和女性细胞。

问题：

（1）在女性中有多少细胞（S）显示巴尔氏体？是否任何一个细胞都有一个以上的巴尔氏体？

（答案略。）

（2）是否任何雄性细胞都有巴尔氏体？有多少？任何雄性细胞都有一个以上的巴尔氏体吗？

（答案略。）

注意事项：

对于典型的女性，在 50%~65% 的口腔细胞中都能发现性染色质。对于典型的男性，只有约 2% 的口腔细胞中可以观察到性染色质。如果在步骤 5 中用 HCl 处理的时间太长，性染色质也有可能被水解。如果处理时间不足够长，一些残留的 RNA 可能与性染色质混淆。

Experiment 8   Observation of the Human Sex Chromatin

I . Experimental Objectives

To observe the human sex chromatin and learn the methods of preparation observation and analysis of chromatin, further to understand the heterochromatin, X chromosome inactivation and other related knowledge.

II . Experimental Principles

In mammals, females carry two copies of all loci on the X chromosome, while males carry only a single copy. One would therefore expect females to express twice as much gene product for sex-linked loci as males. However, a mechanism called dosage compensation prevents this. About 16 days after fertilization, one of the X chromosomes becomes inactivated in all somatic cells. Maternal or paternal X chromosomes are inactivated at random. If an individual is heterozygous for a sex-linked locus, this means that alternate alleles are inactivated in different cells, and that all cells descended from them will express one or the other allele, but not both. Such individuals will be genetic mosaics. Cytologically, the inactivated X chromosome becomes condensed into a small structure in the nucleus called a sex chromatin or Barr body (after one of its discoverers, Dr. Murray L. Barr, a Canadian cytogeneticist). Typical XX females have a single Barr body; typical XY males have no Barr body. In individuals with more than the standard number of X chromosomes, all but one of the X chromosomes become condensed into Barr bodies. Examination of Barr Bodies in hair follicles has been used as "gender test" at the Olympic Games, following suspicions that some eastern European countries attempt to pass off male athletes as females.

III . Experimental Materials

Oral mucosa, compound microscope, dissecting needle, slides and cover slips, forceps, filter paper, and scalpel.

50% acetic acid, alcohol, 2% acetic orcein, HCl, carbol fuchsin.

IV . Experimental Procedures

(1) To reduce the contamination of the epithelial cells from oral bacteria, rinse your mouth several times with tap water.

(2) The slides must be cleaned thoroughly. First wash the slide in detergent, then rinse in distilled water, last rinse in 70% ethyl alcohol and flame the slide to dry.

(3) Gently scrape the lining of your cheek with a toothpick to remove a few epithelial cells. Smear the cells onto the cleaned slide and allow them to air dry.

(4) Drip 95% ethyl alcohol for approximately 2 minutes, drip 70% ethyl alcohol for 2 minutes, drip 50% ethyl alcohol for 2 minutes, and last drip distilled water for 2 minutes.

(5) Transfer the slide to a piece of paper towel. Drip several drops of 5 or 6 mol/ L HCl on the smear for approximately 5 seconds. This should eliminate the RNA but leave the DNA (Time is critical).

(6) Wash the slide with distilled water from a squeeze bottle for 10−15 seconds to dilute the hydrochloric acid over the sink.

(7) Stain the slide in carbol Fuchs (or 2% acetic orcein) for 10−15 minutes at room temperature.

(8) Rinse in distilled water.

(9) Place a cover slip. Locate the cells with low power, and then examine individual cells under high power or oil immersion.

V . Experimental Results

Examine 20 cells from at least one male subject and one female subject.

Questions:

(1) How many cells in the female(s) show Barr bodies? Do any cells show more than one?

(The answer is omitted.)

(2) Do any cells from males show Barr bodies? And if yes, how many do they show? Does any male show more than one?

(The answer is omitted.)

Attention:

In typical females, sex chromatin is found in 50%−65% oral cells. In typical males, the sex chromatin is found in 2% oral cells. If the HCl treatment is too long in step (5), the sex chromatin may also have been hydrolyzed. If not, some RNA may remain and be confused with sex chromatin.

实验九　ABO 血型和人类 Rh 因子

一、实验目的

通过实验掌握人类 ABO 血型的检测方法，进一步理解基因座位（locus）和基因位点（site）的关系，以及复等位基因和基因共显性的概念。

二、实验原理

Landsteiner 在 1990 年发现 ABO 血型系统。他在红细胞上发现 A、B 两种类型的凝集原。血浆中存在两种抗凝集素，分别是抗 A 凝集素和抗 B 凝集素。根据红细胞膜上有无 A、B 凝集原和血浆中有无抗 A、B 凝集素，将人体血型分为 A、B、AB 和 O 四种血型。A 型血只有 A 凝集原和抗 B 凝集素；而 B 型血正好与 A 型血相反。AB 型血包含 A、B 两种凝集原，血清中无抗凝集原。O 型血的血清中含有两种抗凝集素，无凝集原。ABO 血型是由一组复等位基因遗传控制的。

Rh 因子是韦纳在研究恒河猴血液免疫兔子时发现的。一个人的红血细胞表面或者有 Rh 因子或者没有；严格意义上指的是最免疫原性的 D 抗原的血型系统，或 Rh 血型系统，通常称为 Rh 阳性（RH^+ 有 D 抗原）或 Rh 阴性（RH^- 没有 D 抗原）。然而，这种血型系统与其他抗原在临床上是相关的。与 ABO 血型不同，Rh 免疫通常只能发生在输血时或妊娠期间胎盘暴露时。

三、实验材料

血样、采血针、抗血清 A、B 和 Rh、载玻片、棉花、酒精以及一次性注射器。

四、实验步骤

（1）彻底清理载玻片。

（2）把一滴抗血清 A 滴左侧，右侧滴抗血清 B，抗血清 D 在载玻片的中心。

（3）用酒精药棉消毒手指采血部位（一般为无名指末端）。

（4）消毒一次性采血针，用采血针刺手指。

（5）在抗血清的 3 个位置上滴加血液。

（6）用牙签混匀血液和抗凝血清。

（7）两分钟内肉眼观察反应结果，或用最低倍显微镜观察其凝聚的情况，反之可能产生假凝集现象。

五、实验结果

根据下面的描述，观察实验结果。

A 型：血液与 A 血清混合后无凝集现象，与 B 血清混合后有凝集现象。

B 型：血液与 B 血清混合后无凝集现象，与 A 血清混合后有凝集现象。

AB 型：血液与 A、B 两种血清混合后均有凝集现象。

O 型：血液与 A、B 两种血清混合后均无凝集现象。

注意事项：

（1）实验前所用玻片必须清洗干净，以免出现假凝集现象。

（2）血型鉴定时应注意消毒，防止穿刺部位的皮肤感染。

（3）肉眼看不清凝集现象时，应在低倍显微镜下观察。

（4）红细胞悬液及标准血清须新鲜，若受到污染可产生假凝集现象。

思考题：

如果有标准的 A 型红细胞与 B 型红细胞，但无标准血清时，能否进行血型鉴定？

（答案略。）

Experiment 9　　ABO Blood Grouping and Rh Factor in Humans

Ⅰ. Experimental Objectives

Further understand the relationship between the gene locus and the gene loci, multiple alleles and codominant concept of gene through mastering the testing methods of the human ABO blood grouping.

Ⅱ. Experimental Principles

The ABO system of blood types was introduced by Landsteiner in 1900. He found 2 types of antigens present on the RBCs. They are antigen A and antigen B. Similarly, there are 2 types of antibodies present in the plasma called antibody A and B. Based on the presence or absence of antigen and antibodies, human blood is classified into 4 groups of A, B, AB, and O. The A group contains antigen A and antibody B. The B group contains antigen B and antibody A. The AB group contains both antigens A and B but no antibodies. The O group contains no antigen but both antibodies A and B. The ABO blood grouping is inherited by a set of multiple alleles.

Presence of a particular factor is denoted by Rh factor discovered by Weiner in researching the rabbits immunized with the blood of the Macaca rhesus monkey. An individual either has, or does not have, the "Rhesus factor" on the surface of their red blood cells. This term strictly refers only to the most immunogenic D antigen of the Rh blood group system, or the Rh-blood group system. The status is usually indicated by Rh positive (Rh^+ does have the D antigen) or Rh negative (Rh^- does not have the D antigen). However, other antigens are also clinically relevant to this blood group system. In contrast to the ABO blood grouping, immunization against Rh can generally only occur through blood transfusion or placental exposure during pregnancy in women.

III . Experimental Materials

Blood sample; applicator stick; antisera A, B, and Rh; slide; cotton; spirit and disposable injector.

IV . Experimental Procedures

(1) Clean a glass slide thoroughly.

(2) Drip a drop of antiserum A on the left side, antiserum B on the right side and antiserum D at the center of the slide.

(3) Clean the tip of the index finger with cotton soaked in spirit.

(4) Sterilize the disposable injector, then prick the finger tip with the help of a sharp applicator.

(5) Drip a drop of blood in 3 places in the antisera.

(6) Mix the blood and antisera using applicator sticks.

(7) In two minutes results are visible to the naked eye observation, or can be observed by the low power microscope, if over this time false agglutination phenomenon may be produced.

V . Experimental Results

Based on the following describing, observe your experimental results.

Type A: There is no agglutination phenomenon when the blood mixes with A serum except with B serum.

Type B: There is no agglutination phenomenon when the blood mixes with B serum except with A serum.

Type AB: There is the agglutination phenomenon when the blood mixes with both A serum and B serum.

Type O: There is no agglutination phenomenon when the blood mixes with both A serum and B serum.

Attention:

(1) Before starting the experiment the slide and tube must be cleaned to avoid the phenomenon of false agglutination.

(2) Do careful disinfection to avoid puncturing the skin during the identification of blood type.

(3) If you can't view the agglutination phenomenon clearly with naked eyes, use the low power microscope.

(4) The red cell suspension and standard serum must be fresh. If they are contaminated, the phenomenon of false agglutination will be produced.

Question:

If there are standard red blood cells of type A and type B but no standard serum, can you make the identification of blood types?

(The answer is omitted.)

实验十　人类 ABO 血型的群体遗传学分析

一、实验目的

（1）了解 ABO 血型系统人群表型和基因型和它们的遗传。

（2）预测后代的基因频率。

（3）根据 Hardy Weinberg 公式计算基因型的预期比。

二、实验原理

Hardy Weinberg 原理表明，构成一个群体的等位基因频率能一代又一代地保持不变，并在群体中保持平衡，例如，如果不涉及进化，等位基因频率将长期不变。因此进化可以被定义为在长时间内群体等位基因频率的改变。为了使群体的基因构成随时间发生变化，进化力（有时称为进化单元）就必须有打破 Hardy Weinberg 平衡的力量。

ABO 血型系统表明了基因型和表型的关系。它由位于 9 号染色体上的多基因控制，共有三个等位基因，分别为 A、B、O。每一个人都只有两个等位基因（每个基因位点来自每个同源染色体）。A 和 B 基因互为共显性的，共同表达。A 和 B 基因都对 O 基因显性。这样就有四种可能的血型（表型）和基因型，如下表所示。

4 种可能的血型（表型）和基因型

表　型	基因型
Type A	AA or AO
Type B	BB or BO
Type O	OO
Type AB	AB

人类的 ABO 血型遗传的 I^A、I^B、i 基因可能的 6 种基因型频率在 Hardy Weinberg 平衡上由下列三项式决定：

$$\left[p(I^A) + q(I^B) + r(i) \right]^2 = p^2(I^AI^A) + q^2(I^BI^B) + r^2(ii) + 2pq(I^AI^B) + 2pr(I^Ai) + 2qr(I^Bi) = 1$$

由于人类表型为 A 型血与表型为 B 型血者各有 2 种基因型，在实际的抽样观察中不能区分，所以不可能从上述平衡式中求其中任一基因的基因频率。从随机抽样群体的血清学检查中可以观察 4 种血型表型的频率 $\overline{A}$、$\overline{B}$、$\overline{O}$ 及 $\overline{AB}$。由 Hardy Weinberg 平衡定律所预计的，即由理论的表型频率与样本群体中实际观察得到的表型频率的关系式，也即利用下列公式可以分别计算基因 i，I^A 及 I^B 的频率：

$$r(i)^2 = r = O$$

$$p(I^A) = 1 - (q + r) = 1 - \sqrt{(q+r)^2} = 1 - \sqrt{q^2 + 2qr + r^2} = 1 - \sqrt{\overline{B} + \overline{O}}$$

$$q(I^B) = 1 - \sqrt{\overline{A} + \overline{O}}$$

由于许多随机的原因，利用上述公式计算群体中的复等位基因频率时可能出现 $p + q + r \neq 1$ 的情况。如果是这样，则可运用伯恩斯坦（Bernstein）提出的下列公式进行修正：

$$p' = p(1 + D/2), \quad q' = q(1 + D/2), \quad r' = (r + D/2)(1 + D/2)$$

其中，p'，q'，r' 分别为 p，q，r 的修正值，$D = 1 - (p + q + r)$，且有：$p' + q' + r' = (1 + D/2)(1 - D/2) = 1 - 1/4D^2$。

显然，只有当 D 值非常小时，其基因频率之和才非常接近于 1。

三、实验材料

纸笔/计算器和电脑，人的 ABO 血型系统的血清学资料（大学生义务血型鉴定实验资料）。

四、实验步骤

（1）以班级为单位，收集血型鉴定资料。以此作为中国汉族人的一个随机样本（N=2 000），并以此来验证 Hardy-Weinberg 平衡定律。

（2）按收集来的资料，根据 Hardy-Weinberg 定律，把计算所得到的基因频率数据整理在下表中。

中国 ABO 血型的抽样调查表

项目	人　数	A 型	B 型	O 型	AB 型
北方					
南方					
总计					

五、结果分析

（1）按上述公式，可以分别计算出抽样中的南方人、北方人的 4 种血型表型的频率 p、q、r 和 $\overline{A}$、$\overline{B}$、$\overline{O}$ 及 $\overline{AB}$，以及 p'、q' 和 r' 的值。

（答案略。）

（2）根据 Hardy Weinberg 定律，通过本实验的抽样调查具有 $I^A i$、$I^B i$ 和 I 的资料频率来估测汉族人群中所具有的 $I^A i$、$I^B i$ 和 I 基因频率。

（答案略。）

Experiment 10　Population Genetic Analysis of Human ABO Blood Types

Ⅰ. Experimental Objectives

(1) Understand the phenotypes and genotypes in the human ABO blood types system and their inheritance.

(2) Allow predicting the gene frequencies in future generations.

(3) Calculate the expected ratios of the genotypes based on the Hardy-Weinberg formula.

Ⅱ. Experimental Principles

The Hardy-Weinberg Principle states that the frequency of alleles that make up a population will remain constant for generation after generation, so that the population remains in equilibrium, i.e. if it does not evolve the evolution, the allele frequencies do not change over time. Thus, evolution may be defined as a change in allele frequencies in a population over time. In order to make the genetic composition of a population change with time, evolution forces (sometimes referred to as *evolutionary agents*) must disrupt the Hardy-Weinberg equilibrium.

The ABO blood types system aptly demonstrates the relationship between genotype and phenotype. It is a multiple allele system located on chromosome 9. There are three possible alleles of A, B, and O, but of course, each individual has only two alleles (separately on each chromosome of the homologous pair). A and B are codominant, both are expressed in the phenotype. A and B are both dominant to O. Considering these relationships, there are four possible blood types (phenotypes) and their associated genotypes (see from the following table).

Four possible blood types (phenotypes) and their associated genotypes

Phenotypes	Genotypes
Type A	AA or AO
Type B	BB or BO
Type O	OO
Type AB	AB

The 6 possible genotype frequencies of I^A, I^B, i of human ABO blood types may be decided in Hardy Weinberg equilibrium by the following three types:

$$\left[p(I^A)+q(I^B)+r(i)\right]^2 = p^2(I^AI^A)+q^2(I^BI^B)+r^2(ii)+2pq(I^AI^B)+2pr(I^Ai)+2qr(I^Bi)=1$$

Because the type of each A blood type and B blood phenotype in humans is 2 genotypes, it cannot be distinguished actually in the sampling observation and the gene frequency of any gene from the balance population cannot be gotten based on the above formula. The frequency of phenotype of 4 blood types sampled randomly from the population on serological examination can be observed. The 4 blood types are $\overline{A}$, $\overline{B}$, $\overline{O}$ and $\overline{AB}$. The relationship of the phenotype frequencies obtained actually from between the observation and the expected theory by the Hardy Weinberg equilibrium can be used to calculate the genes frequencies of i, I^A and I^B. The formula is as follows:

$$r(i)^2 = r = O$$

$$p(I^A) = 1-\left(q+r\right) = 1-\sqrt{(q+r)^2} = 1-\sqrt{q^2+2qr+r^2} = 1-\sqrt{\overline{B}+\overline{O}}$$

$$q(I^B) = 1-\sqrt{\overline{A}+\overline{O}}$$

With many random causes, $p+q+r\neq1$ may occur when alleles frequencies are calculated using the above formula. If this happens, you can amend it according to the following formula reported by the Bernstein:

$p'=p(1+D/2)$, $q'=q(1+D/2)$, $r'=(r+D/2)(1+D/2)$

Among them, p', q' and r' are the correction values of p, q and r. $D=1-(p+q+r)$ and $p'+q'+r'=(1+D/2)(1-D/2)=1-1/4D^2$.

Obviously, only when the D value is very small, the sum of gene frequencies is very close to 1.

Ⅲ. Experimental Materials

Pen and paper/calculator and computer, the serological data of ABO blood group system (identification information of blood type from the voluntary blood donation of the university students).

Ⅳ. Experimental Procedures

(1) Take the class as a unit, and collect the experiment data of blood groups. This data can be used as a random sample of Chinese Han people (N=2,000) to verify the Hardy-Weinberg equilibrium law.

(2) Using the data collected, calculate the gene frequencies according to the Hardy-Weinberg law, and fill in the following table.

Sampling survey of Chinese ABO blood types

Item	The number	A type	B type	O type	AB type
The North					
The South					
Total					

Ⅴ. Experimental Results

(1) According to the above formula, you can calculate the phenotype frequencies of the 4 blood types sampled from southerners and northerners of which are p, q, r, $\overline{A}$, $\overline{B}$, $\overline{O}$, $\overline{AB}$, p', q' and r'.

(The answer is omitted.)

(2) According to the Hardy Weinberg law, estimate the frequencies of $I^A i$, $I^B i$ and I in Chinese Han population by means of the sampled data of the frequencies of $I^A i$, $I^B i$, and I.

(The answer is omitted.)

实验十一　卡方分析

一、实验目的

卡方检验是用途很广的一种假设检验方法，这里我们主要学习它在资料统计推断中的应用。

二、实验原理

数据统计可以用来确定群体之间是否存在显著差异，或用来预测误差结果。统计检验最常用于检验实验获得的数据是否合适或符合预期。这样的理论数据检验方法就是卡方检验。它可用来测试预期值的偏差是由偶然还是其他因素造成的。

为确定所观察到的数据是否落在可接受的范围内，卡方分析可测试无效假设的真实性，也可用来检验观察到的和预期的数据之间有没有统计上的显著差异。如果卡方分析表明数据差异相差太大，不符合预期的 3∶1，则相反的一个假设被接受。

卡平方计算公式：

$$\chi^2 \quad \sum \frac{(o \quad e)}{}$$

式中，o= 观察到的个体值；

　　　e= 预期的个体值。

自由度（df）由调查的表型数减去 1 得到。如果卡方值数大于表中的临界值，无效假设错误，表明在 0.05 水平上是差异显著的。如果无效假设正确，就意味着只有 5% 的概率你能得到期望的数据。如果在 0.001 水平上被拒绝，就意味着不到 1% 的概率你能得到期望的数据。

三、实验材料

果蝇、过熟的香蕉、广口瓶（1L）、纸巾、橡皮筋、冰袋、培养皿、光学显微镜、冰箱、孵化器。

培养基是一种复杂的混合物，包括琼脂、糖、其他营养补充剂，以及霉菌抑制剂。

四、实验步骤

具体见果蝇的伴性遗传。

举例：练习。

当纯合的长翅果蝇与纯合短翅果蝇交配，F_1 果蝇翅的长度将在中等值。当中等翅果蝇交配之后，可以得到如下结果：230 只长翅的果蝇，510 只中间长翅的果蝇，260 短翅的果蝇。

（1）F_1 中长翅果蝇的基因型是什么？

（答案略。）

（2）写一个描述果蝇翼长度的遗传方式假设。

（答案略。）

（3）根据下列表格，完成相关问题。

习题（3）表

观测性状	期望值 (e)	($o\text{-}e$)	($o\text{-}e$)^2	($o\text{-}e$)$^2/e$
LL	333	-103	10 609	31.86
Ll	666	-156	24 336	36.54
ll	333	-73	5 329	16.00

①上表中的自由度为多少？

有 2 个自由度。

②卡方值为多少？

84.4。

③针对临界值的图表，这些数据的概率值是多少？

小于 0.001。

④根据概率值是接受还是拒绝无效假设？

可以拒绝无效假设，因为卡方值大于表中临界值。

五、结论

从你的实验中推论果蝇翅长的遗传是否符合 1∶2∶1 的比例？在这种情况下，无效假设是否可以接受？

附录：

临界值表

自由度 (*df*)					
1	2	3	4	5	
0.050	3.84	5.99	7.82	9.49	11.1
0.010	6.64	9.21	11.30	13.20	15.1
0.001	10.80	13.80	16.30	18.50	20.5

Experiment 11　　Chi-Square Analysis

Ⅰ. Experimental Objectives:

Chi-square test is a method of hypothesis test with wide application. Here we mainly study its practical application in statistical data.

Ⅱ. Experimental Principles

Statistics can be used to determine if the differences among groups are significant, or to predict the result of error. The statistical tests are most frequently used to determine whether the data we obtain experimentally provides a good suit, or is appropriate for the expected result. This theoretical data testing method is called Chi-square test. Chi-square test can be used to judge whether the deviation from the expected values is accidental error or due to some other circumstances.

To determine if the observed data falls into the acceptable range, Chi-Square analysis is adopted to test the validity of a null hypothesis that there is no statistically significant difference between the observed and the expected data. If the Chi-Square analysis indicates that there exists a significant difference, the records do not meet at variance with the expected 3:1 and an alternative hypothesis is accepted.

The formula of Chi-square is:

$$\chi^2 = \sum \frac{(o-e)^2}{e}$$

Among which, o= observed number of individuals;

e= expected number of individuals.

The df is determined by taking the number of possible phenotypes and subtracting one from it. If the Chi-Square answer is greater than the number from the critical values chart, then the null hypothesis is incorrect. It indicates the difference is significant at 0.05 level. This means only 5 % chance you could get the data expected if the null hypothesis were correct. The probability can also be rejected at 0.001 level, which implies that you have less than 1% chance to get the data expected.

III . Experimental Materials

Drosophila, overripe banana, (1-liter) jar, paper towel, rubber band, ice packs, petri dishes, a light microscope, a refrigerator, an incubator.

The culture medium is a complex mixture of agar, sugars, other nutritional supplements, and mold inhibitors.

IV . Experimental Procedures

See the experiment of the sex-linked inheritance of *Drosophila melanogaster*.

Example: Practice.

When long-winged drosophilas of pure-breeding are mated with short-winged flies of pure-breeding, the F_1 is of an intermediate wing length. When several intermediate-wing-length flies are allowed to interbreed, the results are obtained with about 230 long-winged flies, 510 intermediate-length-winged flies, and 260 short-winged flies.

(1) What is the genotype of the F_1 intermediate wing length flies?

(The answer is omitted.)

(2) Write a hypothesis describing the mode of inheritance of wing length in *Drosophila*.

(The answer is omitted.)

(3) According to the following table, answer the questions.

Observed phenotypes	Expected (*e*)	(*o-e*)	(*o-e*)^2	(*o-e*)^2/ *e*
LL	333	-103	10,609	31.86
Ll	666	-156	24,336	36.54
ll	333	-73	5,329	16.00

① How many degrees of freedom are there according to the above table?

There are 2 degrees of freedom.

② How much is the Chi-Square value?

84.4.

③ Referring to the critical values chart, what is the probability value for these data?

It is less than 0.001.

④ According to the probability value, can you accept or reject the null hypothesis?

I can reject the null hypothesis because the Chi-square value is greater than the critical value according to the table.

V. Experimental Results

From your experiment, infer if the results of inheritance of wing length in Drosophila are close to a 1 : 2 : 1 ratio and if the null hypothesis in this case can be accepted or not.

Appendix:

Critical Values Chart

Degrees of Freedom (*df*)					
1	2	3	4	5	6
0.050	3.84	5.99	7.82	9.49	11.1
0.010	6.64	9.21	11.30	13.20	15.1
0.001	10.80	13.80	16.30	18.50	20.5

实验十二　数量性状的遗传分析

一、实验目的

学习统计分析数量性状遗传试验的数据，估算遗传率和杂种优势表现的程度。了解植物数量性状的遗传规律。

二、实验原理

孟德尔遗传学是离散性状代与代之间遗传传递的定律。例如，孟德尔的实验证明了皱豌豆和光滑豌豆的遗传模式。离散和非连续性状对应的是连续的或数量的性状。植株高度和单株叶数都是连续性状。连续性状可以表示为一个正态分布的连续性状的变化，可绘制成一个钟形曲线。大多数的有关遗传和遗传力讨论都是对连续性状研究而产生的。

数量性状受多基因控制，且各基因间的关系复杂，因此进行数量性状遗传分析时，往往从一对基因（如 A，a）的遗传模型及其基因效应分析着手。现根据加性—显性遗传模型，假设纯合型 AA 和 aa 的加性效应值分别为 d 和 $-d$，中亲值（m）为 $[d+(-d)]/2=0$，由杂合体 Aa 的显性作用所引起的显性偏差为 h，则其基因的作用效应可分解为：

完全显性时：$d=+a$（如果 A 为显性）或 $d=-a$（如果 a 为显性）。

超显性时：$d>+a$（A 对 a 超显性）；$d<-a$（a 对 A 超显性）。

部分显性时：$-d<h<d$。

群体的表现型方差标记为 V_P，V_P 通常为所有的遗传方差之和，V_E 为环境方差，则群体的表现型方差（V_P）应为基因型方差（V_G）和环境方差（V_E）之和。

$$V_P = V_G + V_E$$

$$V_G = V_A + V_D$$
$$V_P = V_A + V_D + V_E$$

一般意义上的遗传力又叫广义遗传力 (h_b^2)，是指遗传方差占表型方差的比例 ($h_b^2 = V_G / V_P$)。狭义遗传力 (h^2) 是加性方差占表型方差的比例 ($h^2 = V_A / V_P$)。

F_2、$B_1(F_1 \times P_1)$、$B_2(F_1 \times P_2)$ 群体的方差组成分析为：

$$V_{F2} = 1/2D + 1/4H + V_E$$
$$V_{B1} + V_{B2} = 1/2D + 1/2H + 2V_E$$

式中，$D = \Sigma d^2$，是各基因加性效应方差的总和；$H = \Sigma h^2$ 是各基因显性偏差方差的总和。根据上述群体方差的组成分析，可统计分析数量性状遗传试验的数据，并按公式估算遗传率。杂种优势是指杂种一代（F_1）在产量、品质、生活力、生长势、抗逆性和适应性等方面比其双亲优越的现象。

三、实验材料

将 P_1（长果穗）、P_2（短果穗）及其杂种后代 F_1、F_2，回交后代 B_1、B_2 以及对照种的穗于同年种植，分别收获后，按世代测量、记录果穗长度。

计算器、尺子和托盘天平。

四、实验步骤

1. 基本参数的计算

（1）计算各世代果穗长度的平均数、方差（V）及标准差（S）。

（2）计算环境方差（V_E）。F_2 的环境方差采用下列方法估算：

$$V_E = 1/3(V_{P1} + V_{P2} + V_{F1})$$

2. 遗传率的估算

（1）广义遗传率。
$$h_b^2 = V_G / V_P = \left[\left(V_{P2} - 1/3(V_{P1} + V_{P2} + V_{F1}) \right) / V_{P2} \right] \times 100\%$$

（2）狭义遗传率。
$$h^2 = V_A / V_P = \left[\left(2V_{P2} - (V_{B1} + V_{B2}) \right) / V_{P2} \right] \times 100\%$$

五、实验结果

分别观察比较亲本、F_1 和对照相互之间在性状上的差异。

（1）分别调查上述群体的穗长、穗重，并将结果记录下来。

（2）根据调查结果，按照公式，分别计算 F_1 代的平均优势、超亲优势、竞争优势和优势指数等。

Experiment 12　　Genetic Analysis of Quantitative Traits

Ⅰ. Experimental Objectives

Learning and analyzing the data of quantitative traits in genetic tests, and then estimating the ratio of inheritability and the expression degree of heterosis. Understanding the law of inheritance of quantitative traits in plants.

Ⅱ. Experimental Principles

Mendelian genetics provides the laws that govern the passing on of discrete traits from one generation to the next. For example, Mendel experimentally demonstrated particular patterns of inheritance for smooth and wrinkle peas in a population of pea plants. Discrete or discontinuous traits are contrary to the continuous or quantitative traits. Height and leaf number in plants are continuous traits. Continuous traits can be represented as a normal distribution of continuous change, and then normal distribution is graphed as a bell curve. Most philosophical discussions about heredity and heritability arise from the study of continuous traits.

Quantitative trait is most controlled by multiple genes, and the relationship among the different genes is complex. We often analyze genetic models and gene effects from a pair of genes (such as A, a) of quantitative traits. According to the additive-dominant genetic model, we hypothesize that the values of additive effect of homozygous AA and aa are d and $-d$, mid-parent value (m) is $[d+(-d)]/2=0$ and the value dominant deviation caused by the dominant role of heterozygote Aa is h. Then the effect of the gene can be decomposed into the following formula.

Complete dominance: $d = +a$ (if A is dominant) or $d = -a$ (if a is dominant).

Over dominance: $d > +a$ (A is dominant over a); $d < -a$ (a is dominant over A).

Partial dominance: $-d < h < d$.

Variance in phenotype is symbolized as V_P. V_P is actually the sum of all the genetic variance. V_E is the environmental variance. Phenotypic variance (V_P) is the sum of the genotypic variance (V_G) and environmental variance (V_E):

$$V_P = V_G + V_E$$
$$V_G = V_A + V_D$$
$$V_P = V_A + V_D + V_E$$

General sense heritability is called broad sense heritability (h_b^2) which refers to the proportion of genetic variance accounted for phenotypic variance($h_b^2 = V_G/V_P$). Narrow sense heritability (h^2) refers to the proportion of additive genetic variance accounted for phenotypic variance ($h^2 = V_A/V_P$). The components of variance analysis in F_2, $B_1(F_1 \times P_1)$ and $B_2(F_1 \times P_2)$ populations are as follows:

$$V_{F2} = 1/2D + 1/4H + V_E$$
$$V_{B1} + V_{B2} = 1/2D + 1/2H + 2V_E$$

In the above formula, $D = \Sigma d^2$ is the sum of additive effect variance; $H = \Sigma h^2$ is the sum of dominant error variance of each gene. According to the above variance of components analysis, we can analyze the quantitative traits in genetic tests and then estimate the heritability based on the formula. Heterosis is the phenomenon that a hybrid (F_1) on the yield, quality, life stress, growth, resistance and adaptability is superior to its parents.

III . Experimental Materials

P_1 (long ear), P_2 (short ear), hybrids F_1 and F_2, backcross progenies B_1 and B_2, and ear control all of which are planted in the same year, and harvested separately and the ear lengths are recorded according to the generation respectively.

Calculator, a ruler, and tray balance.

IV . Experimental Procedures

1. Calculation of basic parameters

(1) Calculate the average number, variance of ear length (V) and standard deviation (S) of each generation.

(2) Calculate the environmental variance (V_E). The V_E of F_2 can be estimated according to the following methods:

$$V_E = 1/3(V_{P1} + V_{P2} + V_{F1})$$

2. Estimation of heritability

(1) Broad sense heritability.

$$h_b^2 = V_G / V_P = \left[\left(V_{P2} - 1/3(V_{P1} + V_{P2} + V_{F1}) \right) / V_{P2} \right] \times 100\%$$

(2) Narrow sense heritability.

$$h^2 = V_A / V_P = \left[\left(2V_{P2} - (V_{B1} + V_{B2}) \right) / V_{P2} \right] \times 100\%$$

V. Experimental Results

Observe and compare the differences in character among the parental, F_1 and the control.

(1) Investigate the ear length, the ear weight in the above-mentioned population, and record the results.

(2) According to the survey results and the formula, calculate the average heterosis, the superior heterosis, the competitive heterosis, and the heterosis index in F_1 generation.

第三部分

实验十三　孟德尔遗传实验

一、实验目的

在本实验中，我们将实验孟德尔遗传以理解孟德尔的三大定律，包括分离定律、独立分配定律和显性定律。

二、实验原理

孟德尔遗传的原理最早源自 19 世纪奥地利僧侣格雷戈尔·约翰恩·孟德尔，并以其命名。他用豌豆植物杂交实验的结果阐述出自己的观点。孟德尔栽培并实验了大约 5 000 株豌豆植物。从这些实验中，他归纳出两个普遍化的结论，即后来大家所知的孟德尔遗传原理或孟德尔遗传。

三、实验材料

本次实验使用的材料是不同颜色的小球，以此来检验孟德尔定律。每种颜色将被视作不同的遗传特征。

四、实验步骤

各色小球代表常染色体显性和常染色体隐性。

在本实验中，黄色球代表常染色体显性性状的黄粒豌豆，绿色球代表常染色体隐性性状的绿粒豌豆。将两种球（黄色和绿色）混合后放在两个不透明的袋子里，并且每次随机地从每个袋子里取一个球。将结果记录下来。此实验重复 100 次。

五、实验结果

根据孟德尔定律，预计 75% 的结果显示黄色表型，25% 显示绿色表型。最终得到的结果采用卡方检验进行分析。

六、思考题

（1）为什么孟德尔用豌豆做实验？
（2）实验结果能否通过卡方检验得到证实？

Experiment 13　Mendelian Inheritance

I. Experimental Purposes

In the present experiment, the Mendelian inheritance will be examined to understand the three Mendelian laws, including Law of Segregation, Law of Independent Assortment, and Law of Dominance.

II. Principles

The principles of Mendelian inheritance were named after and first derived by Gregor Johann Mendel, a nineteenth-century Austrian monk who formulated his ideas after conducting hybridization experiments with pea plants. Mendel cultivated and tested some 5,000 pea plants. From these experiments, he induced two generalizations which later became known as *Mendel's Principles of Heredity* or *Mendelian Inheritance*.

III. Materials

The materials used in this experiment are small balls of different colors to exam the Mendelian laws. Different colors represent different genetic traits.

IV. Procedures

The small balls of different colors show autosomal dominance and autosomal recessiveness.

In this experiment, yellow balls represent the autosomal dominant trait for yellow seed peas and green balls represent the autosomal recessive trait for green seed peas. Mix the balls (yellow and green) and put them in two non-transparent bags from which one ball will be taken randomly each time. The results will be recorded and the experiment will be repeated for 100 times.

V. Results

According to Mendelian law, it is expected that 75% of the results show Yellow phenotype and 25% show Green phenotype. The final results will be analyzed by using Chi-square Test.

VI. Questions

(1) Why did Mendel use pea plants for his experiment?

(2) Can the results of the experiment be confirmed by Chi-square test?

实验十四　植物多倍体的诱发及鉴定

一、实验目的

了解人工诱导多倍体的原理及一般方法。

二、实验原理

多倍体植物是指每个细胞中的染色体数具有2套或更多套数的植物。许多植物都是多倍体。可育多倍体植物和二倍体相比形体较大，生产性也较好。因此育种工作者常利用杂交或其他方式来生产多倍体植物。在自然界，多倍体植物可以自发产生，其产生机制包括有丝分裂和减数分裂失败，以及未减数分裂配子融合等。

在野生和驯化的植物物种中都可以发现同源多倍体（如马铃薯）和异源多倍体（如油菜、小麦、棉花）。大多数多倍体与亲本相比具有杂种优势，并表现在其形态变化上，这些形态变化有助于物种形成和新的生态位开拓。导致异源多倍体生物形态变化的可能机制包括基因剂量效应（从基因组含量多拷贝），不同层次的基因调控层次联合，染色体重排，以及表观遗传模型等，所有这些因素都影响基因的含量和/或表达水平。一些植物如蒲公英是通过无性繁殖，也可以通过单性生殖或其他无性方式繁殖。其他多倍体植物是可育的，例如硬粒小麦（*Triticum turgidum durum*）用来做面食，是四倍体（具有四套染色体），而面包小麦（*Triticum aestivum*）是六倍体（具有六套染色体）。

秋水仙碱是一种生物碱，可从秋水仙被子植物中得到，其属于秋水仙科。秋水仙碱是一种剧毒生物碱，长期接触可导致白细胞异常。该化合物阻止微管蛋白纤维的形成，也阻止了一些细胞运动和白细胞的迁移，其是细胞分裂必需的。秋水仙碱的副作用可能包括腹泻、腹胀和胀气。秋水仙碱是诱变剂，它阻碍了纺锤体的形成。秋水仙碱还被用于植物育种产生多倍体株。秋水仙素诱导多倍体，可引起植株形态学、组织学和细胞学上的变化，甚至在基因表达水平上的变化。

三、实验材料

洋葱，植物种子或幼苗。显微镜、酒精灯、水浴锅、培养皿、镊子、剪刀、解剖针、刀片、载片以及盖片。

0.2%~0.4% 秋水仙素水溶液、卡诺氏固定液、1% 醋酸洋红、70% 酒精、0.1%~0.2% 升汞和蒸馏水。

四、实验步骤

1. 洋葱材料的处理

将秋水仙素溶液倒入小培养皿中，放上洋葱鳞茎，使其生根部位刚好和液面接触。同时另一培养皿内放置清水，放洋葱鳞茎作为对照。

在 25℃ 下培养数日，待鳞茎长出幼根时即可进行观察。

经加倍的根尖都较正常对照的肥大。用刀片切取经处理而肥大的根尖及对照的根尖（长 2~5 mm），投入卡诺氏固定液中固定。按前述的方法进行染色体制片，计数染色体数目的变化。

2. 处理种子

这种方法适用于发芽快或能在数天内发芽的种子。

先将植物种子洗净，用水浸一天或干燥种子用 0.1%~0.2% 升汞溶液消毒 8~10 min，再用清水洗净。

然后将其摆放在铺有湿滤纸的一些培养皿中，在其中一部分培养皿中加入 0.2% 秋水仙素溶液，在另一部分培养皿中注入清水作为对照，并将准备好的种子种在里面。

为了避免溶液蒸发，可在培养皿上加盖，置于培养箱中保持 25℃ 左右使种子发芽。

种子萌发后，应继续处理 24 h。在处理过程中，注意溶液的蒸发，随时添加清水，以保持原处理药液浓度。

处理后，用清水冲洗净种子上的残留物质，再播种。处理适度的种子比对照的发芽稍慢，种芽胀大。从形态上可初步区分出是否加倍成功。

3.处理幼苗或成株

对于发芽迟缓的种子，在其出苗后处理幼苗效果更好。将蘸有 0.1%~0.4% 秋水仙素的棉球，置放于顶芽、腋芽的生长点处，并且经常滴加清水以保持药液浓度。处理幼苗或成株的生长点所需时间在 24~28 h，处理后植株将进一步生长，我们要进行观察和鉴定。

五、实验结果

1.植物形态观察

在植株整个生长期形态观察都可以进行。每隔一定的时期可以观察种子生长情况，并将这些形态变化记录下来。

2.生理变化

对秋水仙素处理后的植株和未处理的植株的各种生理参数进行调查（气孔密度和长度）。取其叶片，用二甲苯处理下表皮，使其保持干燥。然后可以用镊子来去掉下表皮，用蓝色染料将其染色，在 20 倍显微镜下观察叶面气孔密度。在形态观察的基础上，可进一步进行镜检以观察染色体数目的变化。

问题：

写出本组诱发植物多倍体的方法，以及鉴定方法。

（答案略。）

附：0.2%~0.4% 秋水仙素的配制：

取秋水仙素 1 g（先用少许 95% 酒精助溶），溶于 250~500 mL 蒸馏水中。亦可制成较高浓度的母液，放入棕色玻璃瓶内，用时再稀释到所需的浓度。

Experiment 14   Induction and Identification of Plant Polyploid

I . Experimental Objectives

Understand the principles and general methods of the induction of artificial multiploid.

II . Experimental Principles

Polyploid refers to plants that have more than two complete sets of chromosomes. Many plants are polyploid. Polyploid plants, if viable, are often larger or more productive than diploid plants, and plant breeders often deliberately produce such plants by crossing species or other means. Polyploid plants can arise spontaneously in nature by several mechanisms, including meiotic or mitotic failures, and fusion of unreduced ($2n$) gametes.

Both autopolyploids (e.g. potato) and allopolyploids (e.g. canola, wheat, and cotton) can be found among both wild and domesticated plant species. Most polyploids display heterosis relative to their parental species, and may display novel variation or morphologies that may contribute to the processes of speciation and eco-niche exploitation. The mechanisms leading to novel variation in newly formed allopolyploids may include gene dosage effects (resulting from more numerous copies of genome content), the reunion of divergent gene regulatory hierarchies, chromosomal rearrangements, and epigenetic remodeling, all of which affect gene content and/or expression levels. Some plants such as dandelions are sterile, but can be reproduced by apomixis or other asexual means. Other polyploid plants are fertile. For example, durum wheat (*Triticum turgidum durum*), which is used to make pasta, is tetraploid (it has four sets of chromosomes), while bread wheat (*Triticum aestivum*) is hexaploid (it has six sets of chromosomes).

Colchicine is an alkaloid which is obtained from Colchicum autumnale of angiosperm and belongs to the family of Colchicaceae. Colchicine is an extremely toxic alkaloid. Long-term touching can result in white blood cell anomalies. A compound, blocking the assembly of microtubules-protein fibers and some kinds of cell movements and neutrophil migration, is necessary for cell division. Its side effects may include diarrhea, abdominal bloating and flatulence. Colchicine is also considered as a mutagenic agent since it hinders the spindle formation. Colchicine is also used in plant breeding to produce polyploid strains. The effects of artificial polyploidial caused by colchicines may induce some changes in morphology, cytology, histology and even in gene expression level.

III. Experimental Materials

Onion, plant seeds or seedlings. Microscope, alcohol lamp, water bath, petri dishes, tweezers, scissors, dissecting needle, blade, slide, cover plate.

0.2%−0.4% colchicine solution, Kano's fixative solution, 1% carmine acetate, 70% alcohol, 0.1%−0.2% $HgCl_2$ and distilled water.

IV. Experimental Procedures

1. Treatment of onion material

Pour colchicine solution into a small dish where the onion bulbs are treated. The roots of onion bulbs need to contact the surface of the liquid. At the same time, take another plant onion bulb and petri dish and add water as control.

Culture the onion bulbs at 25 ℃ for several days. Observe when the bulbs grow young roots.

The root tips of colchicine treated plants are bigger in comparison with the controlled plants. Cut the root tips (2−5mm) of treatment and control with a knife and put them into the Carnoy's fixative solution. Count the changes in the number of chromosomes according to the previous method.

2. Seeds treatment

This method is suitable for the seeds with fast germination, or germination within a few days.

Wash the seeds and immerse them into water for one day or sterilize the dry seeds with 0.1%–0.2% $HgCl_2$ for 8–10 minutes and then rinse with water.

Then place a wet filter paper inside of all the petri dishes. Add 0.2% colchicine solution into some petri dishes and add water into the other dishes as the control. Plant the seeds prepared into these petri dishes.

The lid can be used to avoid evaporation. Place them into incubator at 25°C until the seeds germinate.

After germination, keep them in incubator for another 24 hours. During the process, we still need to add some water to keep the original concentration because of liquid evaporation.

After treatment, wash the residual from the seeds and sow them into pots. The seeds treated germinate slightly slower than the seeds under the control and its buds swell. You can identify successfully if they are polyploids or not from the shapes of the buds.

3. The treatment of seedlings or adult

It will have better effects to treat the seedlings after germination for the seeds are retarded in germination. Place the cotton balls dipped with 0.1%–0.4% colchicine on the growth points or apical buds, and often drip water on them to maintain the same liquid concentration. The time of treatment of seedlings or growth points is form about 24 to 28 hours. Observe and identify when plants further grow.

V . Experimental Results

1. Plant morphology observation

Plant morphological investigation can be carried out during the whole growth period. The potted seeds are observed at regular intervals and the observations (changes in the plant's morphology) are recorded.

2. Physiological changes

Various physiological parameters (stomatal density and stomatal length) are examined between the control plants and the colchicine treated plants. Fresh leaves of the control and colchicine treated plants are taken and xylene is applied to the lower epidermis and kept for drying. After drying, the lower epidermis is peeled off with forceps and is stained with trypan blue for visualization of stomatal density under light microscope at 20 times of magnification. Based on the observation of morphology, you can further observe the changes in chromosome number under the microscope.

Question:

Write the method with which polyploid plants are induced and how to identify if they are polyploid or not.

Appendix:

Preparation for 0.2%−0.4% colchicine:

Take 1g colchicine (help to dissolve it with a little 95% alcohol), and dissolve it in 250−500 mL of distilled water. A higher concentration of stock liquor can also be made and stored in a brown glass bottle. Dilute to the required concentration before using.

实验十五　植物有性杂交技术

一、实验目的

（1）理解水稻或玉米有性杂交的原理。

（2）了解水稻或玉米的花器构造、开花习性、授粉、受精等有性杂交基础知识。

（3）掌握水稻或玉米的有性杂交技术。

二、实验原理

杂交是最常用的植物育种技术。杂交种也被称为 F_1，是两个不同的亲本杂交的后代。其杂交种结合父母双方优良性状的现象被称为"杂种优势"，表现在幼苗高度上和更好的生产性。这些比普通父母本表现好的高产品系，也被称为自交系。为了保持杂种优势，原有的父母本每一年都得杂交。很少有人从杂交种上保留种子做商业品种的，虽然在育种上经常这样做。杂交可以正交，也可反交。例如，♂A×♀B 和 ♀B×♂A。如果杂交种与其亲本之一杂交，称为回交。测交就是研究隐性性状的回交。远缘杂交包括种间杂交，其父母本来自两个不同的属种间；种内杂交，其父母来自两个不同的品种。

水稻的"花"叫颖花。花器官里面是外稃和内稃。颖花的组成见本文后所附图片。每个颖花里有 6 个雄蕊。每个雄蕊由一个花药和花丝组成。花药包括 4 个细长的储存花粉的囊，细长花丝是花药主干。花丝中的维管束用于给花药输送养分和水分。水稻的心皮由部分雌花构成——柱头和子房。柱头接受花粉颗粒，然后将其输送到子房中受精。外稃比内稃相对较大。当颖花关闭后外稃部分封闭内稃，外稃尖的一端被称为芒。

玉米是业余植物育种实验的一个很好的植物材料，因为它容易操作，并且在 F_1 代籽粒上经常呈现出可见的变化。玉米具有不完全花，雄花和雌花在同一植物上（见本文后附的图1和图2）。雄蕊是花穗，雌花是玉米穗。覆盖玉米穗表面且可授粉的花丝是玉米的柱头。

三、实验材料

普通水稻与玉米、品种3~4个、镊子、剪刀、玻璃透明纸袋、回形针、大头针、铅笔、小纸牌（白色，大小为3 cm×4 cm）、放大镜、棉花球、六磅热水瓶、500 mL烧杯以及95%酒精。

四、实验步骤

（一）水稻有性杂交过程

1. 去雄

（1）剪颖法。在授粉前一天下午，用剪刀斜剪去所需的母本穗的顶部和底部1/3的穗花，留下中间的部分。用剪刀斜角剪去颖花顶部1/3颖壳。用镊子取出六个雄蕊，去雄必须小心，不能损伤雌蕊。为防止外来花粉的污染，去雄后套袋。

（2）温汤去雄法。温汤去雄法可用于剪颖法的延伸，选择离开花还有3~4天的颖花做母本。大约在开花一个小时之前，去掉开过的和未发育好的颖花，把稻穗轻轻压弯（小心避免断裂），颖花全部浸入40℃~44℃的温水之中5~10 min。这种处理可使颖花正常开花，避免损伤。然后用镊子取出六个雄蕊。

2. 杂交

（1）第二天早晨（通常在上午九时），取所需的开花父本颖花。

（2）剪开原本插在去雄母本顶部的纸袋，使颖花暴露。

（3）将父本颖花插入已剪开的纸袋，上下抖动使花粉散开。

（4）然后用别针将纸袋口封住。

（5）最后把彩色线系在穗的基部来作为标识。

（二）玉米杂交程序

（1）决定需要杂交的玉米品种，种植到试验地中。

（2）因为玉米穗在较短的时间内就散完粉，所以为了在同一时间有不同品种植物的花粉，在一定时期每个品种每周播种一次。

（3）在玉米抽雄和出丝之前，在植株顶部和玉米穗上套袋并用夹子固定。

（4）当穗丝可见时准备授粉，提前 1 天给所选的玉米雄花套袋。

（5）套袋雄花的纸袋基部应牢牢固定以防止花粉脱落和受其他花粉的污染。

（6）第二天早上用力摇纸袋，使花粉松落。弯曲玉米天花，以防止花粉掉出，取下纸袋。

（7）移去花丝上的纸袋，把从天花上取下的纸袋放在穗上，小心不要抖落任何花粉。

（8）将盛有花粉的纸袋套在玉米穗上。

（9）给已授粉的玉米穗编号标记。

（10）纸袋留在玉米穗上直到叶片干枯。最后收获籽粒并保存以便下次种植。

五、实验结果

授粉后就该挂标签了。把标签挂在茎秆下方已授粉的花上。标准的标签写法如下：每个授粉的花应挂上标签，标签的内容包括种子母本的名字、字母 X（指杂交）、父本的名字和杂交日期。杂交种子标记后，就该在笔记本上记录所做的杂交和自交。保持完整和准确的育种记录是非常重要的。

图 1　玉米雄花

图 2　玉米雌花

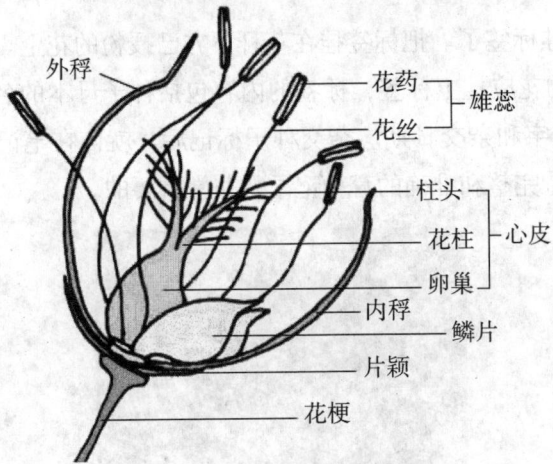

图 3　水稻花的模式

Experiment 15   Sexual Hybridization of Plants

I . Experimental Objectives

(1) Understand the principles of the sexual hybridization in rice or corn.

(2) Obtain the basic knowledge of flower structure, pollination, fertilization and other sexual hybridization in rice or corn.

(3) Master the techniques of the sexual hybridization in rice or corn.

II . Experimental Principles

Hybridization is the most frequently employed plant breeding technique. A hybrid variety, also referred to as the F_1, is the direct product of crossing two genetically different parents. In hybrids, the positive qualities of both parents are combined, resulting in a phenomenon called "hybrid vigor" or "heterosis" with young seedlings being higher in height and the mature plant having better reproductive characteristics. These factors result in higher yields than ordinary parents, also called inbreds. Therefore, in order to maintain all the desirable characteristics of the hybrid, the original parents are crossed each year. It is seldom worthwhile to save seeds from the hybrid for commercial production, although it can be used in a breeding program. Crossings may be direct or reciprocal. For example, the hybrids ♂A × ♀B and ♀B × ♂A are reciprocal. A hybrid crossing with one of the parent is called a backcross. A testcross involves a hybrid backcrossing with a parent with recessive trait studied. Distant hybridization includes intergeneric hybridization—crossing between the two different genera and intraspecific hybridization—crossing between two different species.

The rice "flower" is called a SPIKELET. Inside the floral organs are the lemma and the palea. A spikelet consists of all the parts shown in the attached figures at the end of this article. There are 6 stamens in each rice flower. Each stamen is composed of an anther and a filament. An anther contains 4 elongated sacs where pollen grains are stored. The filament is a long, thin stem that holds the anther. The vascular bundles in the filament transport nutrients and water to the anther. The carpel consists of the female parts of the rice flower—the stigma, and the ovary. The stigma receives pollen grains, which then are transported into the ovary, where fertilization occurs. The lemma is relatively larger than the palea. When the spikelet is closed, the lemma partly encloses the palea. The pointed end on top of the lemma is called an awn.

Corn is a good plant for the amateur plant breeder to experiment because it is easy for operation and often shows visible changes in the kernels of F_1 generation. Corn has an imperfect flower, with both staminate and pistillate flowers on the same plant (See the attached figures 1 and 2 at the end of this article). The staminate flower is the tassel, and the pistillate flower is the ear. The silk, which has a hairy surface throughout most of its length that is receptive to pollen, is the corn's stigma.

III . Experimental Materials

Ordinary rice and maize, 3–4 varieties, operation forceps, scissors, transparent glass paper bag, paper clips, pins, pencil, a small card (white, size 3 cm × 4 cm), a magnifying glass, cotton balls, six-pound thermos bottle, 500 mL beaker and 95% alcohol solution.

IV . Experimental Procedures

Part 1 Procedures of sexual hybridization in rice

1. Emasculation

(1) Clipping method.

On the evening prior to the experiment, the top 1/3rd and bottom 1/3rd portions in the panicle of the desired female parent are clipped off by using scissors, leaving the middle spikelets. With the help of scissors again, the top 1/3 portion in each spikelet is clipped off in a slanting position. Six anthers present in each spikelet are removed with the help of the needle (Emasculation). Care must be taken during emasculation not to damage the gynoecium. Then prevent contamination from the foreign pollen, the emasculated spikelets are covered with a butter paper bag.

(2) Hot water method.

A method of hot water emasculation is used for the extension of the clipping method. Panicles before 3 or 4 days of blooming are chosen as female parents. An hour or so before blooming, the panicle is selected and underdeveloped and opened spikelets are removed. Now, the tiller is bent over (be careful to avoid breaking) and the selected panicle is immersed in hot water (40 °C −44 °C) contained in a thermos bottle for 5 to 10 minutes. This treatment causes the florets to open in a normal manner and also avoids injury. Then, emasculation is done by removing the six stamens with fine forceps or needles.

2.Crossing

(1) On the next day morning (usually at 9.00 a.m.), the bloomed panicle from the desired male parent is taken.

(2) The top portion of the butter paper bag, which was originally inserted in the emasculated female parent, is now cut to expose the panicle.

(3) The male parent panicle is inserted in an inverted position into the butter paper bag and stunned in both ways in order to disperse the pollen.

(4) The opened butter paper bag is closed using a pin.

(5) Colored thread may be tied at the base of the panicle to identify the crossed ones.

Part 2 Procedures of sexual hybridization in corn

(1) Determine the varieties of corn that you wish to cross and plant the seeds in your experimental area.

(2) Corn tassels lose their pollen in a relatively short time, so in order to get the pollen of the different varieties ready at about the same time, plant a few seeds of each variety weekly over a period of weeks.

(3) As soon as the plants develop ear shoots and before the silks emerge, cover the shoots with a loosely fitted bag secured at the bottom with a paper clip or string.

(4) When the silks are visible, the ear is ready for pollination. The appropriate tassel should then be bagged for use on the next morning.

(5) The bag should be secured tightly at the base of the tassel to keep the pollen from falling out and to keep it from becoming contaminated with other pollen.

(6) The following morning shake the bagged tassel vigorously to loosen the pollen. Remove the bag from the tassel, bending the tassel downward to prevent pollen from spilling out.

(7) Remove the bag covering the silk. Place the bag from the tassel over the ear; Take carefully not to spill any pollen.

(8) Place the bag containing pollen on the ear of the corn.

(9) Label each of the ears that have been pollinated with the name or number of the plants that are serving as seeds.

(10) Leave the bagged ear on the plant until the leaves dry. Then harvest the kernels and save them for future planting.

V. Experimental Results

Immediately after pollination, it is time to label the seed parents. Attach the label to the stem just below the flower that has been pollinated. The standard method of labeling is as follows: each pollinated flower should bear a label containing the name of the seed parent, the letter X (to signify a cross), the name of the pollen parent, and the date of the cross. Once the seed parent is labeled, your next step is to record the cross or self in a notebook. Keeping complete and accurate records of your breeding operations is very important.

Figure 1 Male inflorescence

Figure 2   Female inflorescence with young silk

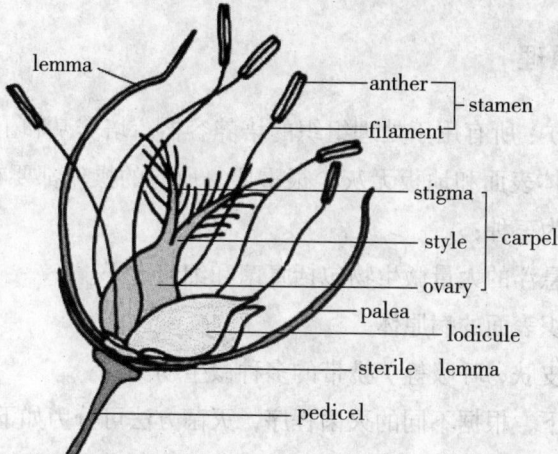

Figure 3   Pattern of rice floral

实验十六　　灭菌操作技术

一、实验目的

无菌培养技术的建立和维护。

二、实验原理

无菌环境维护：所有用于处理组织的培养容器、培养基和工具以及外植体必须消毒。保持物体表面和地板无灰尘很重要。所有的操作都要在消毒柜中进行。污染的途径可分为三种：

（1）空气中悬浮的大量微生物包括真菌和细菌孢子。

（2）植物组织表面的病原体。

（3）人体（皮肤、呼吸等）携带的多种微生物。

在一般情况下，根据不同的灭菌程序，灭菌方法可分为如下几种：

（1）高压蒸汽灭菌（无菌培养基制备、培养容器和工具）。

（2）过滤灭菌（无菌植物生长调节剂的制备）。

（3）外植体（分离的组织）的消毒使用化学灭菌剂，如氯化汞和次氯酸钠。

三、实验材料

灭菌锅、水、瓶、超净工作台、各种培养器皿、培养基、烘箱、酒精灯，以及细菌过滤器。

四、实验步骤

1. 蒸汽或湿灭菌（高压）

灭菌效果取决于压力锅的蒸汽压力。所用设备的尺寸可以小到 1 升或大到几千升。大多数器皿和培养基在高压蒸汽灭菌时的温度一般在 115℃~135℃。一般标准灭菌温度为 121℃，压力为 15 磅（磅每平方英寸），时间为 15 min。这个温度是杀死嗜热微生物的必要温度。灭菌时间和温度取决于灭菌液体的体积和器皿的厚度。压力一般控制在 1 kg/cm^2，即 1.034×10^5Pa，灭菌 20 min。灭菌结束，关闭热源。当压力降至 0 时，应立即开启锅盖，取出物品。

2. 过滤灭菌

一些生长调节剂如氨基酸和维生素是热不稳定的，在高压灭菌下会被破坏。因此这些试剂通过 0.22~0.45 μm 过滤膜过滤灭菌。

3. 辐射

辐射是在紫外线辐射条件下进行的，常被限制用在耗材如培养皿和移液管等内。紫外线灯也可用于杀死在工作台上、房间内的微生物。在工作时，因紫外线灯对人有危害，所以应当关闭。紫外线对皮肤和眼睛可造成伤害。

4. 无菌操作台

这是用于无菌操作的主要设备。台内有从后至前的水平气流。电风扇吸入的空气通过粗过滤器，然后再通过细菌过滤器（HEPA）。高效微粒空气滤波器的空气是直线流通的。小心不要太多打扰空气的流动性。在开始任何实验前，需要用 70% 酒精清洁表面。空气过滤器应定期清洗、更换。

五、实验结果

可取少数培养基置于 37℃恒温箱内 24 h，若无菌生长可保存备用。斜面培养基从锅内取出趁热摆成斜面，凝固后再置于 37℃恒温箱内 24 h，若无菌生长即可保存备用。

思考题：

（1）为什么灭菌器内需要灭菌的物品不能摆得太密？

（答案略。）

（2）灭菌的温度是多少？灭菌结束后要注意什么？

（答案略。）

Experiment 16　　Aseptic Culture Techniques

I . Experimental Objectives

The establishment and maintenance of aseptic culture techniques.

II . Experimental Principles

Maintenance of aseptic environment: All culture vessels, media and instruments used in handling tissues as well as the explants must be sterilized. The importance is to keep the things surface and floor free of dust. All operations are carried out in laminar air-flow of a sterile cabinet. Infection can be classified in three ways:

(1) The air contains a large number of suspended microorganisms in the form of fungal and bacterial spores.

(2) The plant tissue is covered with pathogens on its surface.

(3) The human body (the skin, breathe, etc) carries several microorganisms.

In general, the methods of sterilization can be grouped as follows under different sterilization procedures.

(1) Sterilization is done in autoclave (Preparation of sterile media, culture vessels and instruments).

(2) Filter sterilization (Preparation of sterile plant growth regulators).

(3) Explants (isolated tissues) are sterilized with chemical sterilants, e.g. $HgCl_2$ and NaClO.

III . Experimental Materials

Autoclave, water, bottles, laminar airflow cabinet, all types of culture vessels, media, oven, alcohol lamp, and filtering bacteria filter.

IV. Experimental Procedures

1. Steam or wet sterilization (autoclaving)

The sterilization effect relies on the pressure in a domestic pressure cooker. The size of the equipment used can be as small as one liter or even as large as several thousand liters. Most instruments and nutrient media are sterilized by an autoclave which has a temperature range of $115°C - 135°C$. The standard conditions for autoclaving are a temperature of $121°C$ and a pressure of 15 psi for 15 minutes to achieve sterility. This figure is based on the conditions necessary to kill thermophilic microorganisms. The time taken for liquids to reach this temperature depends on their volume. It may also depend on the thickness of the vessel. The pressure is generally controlled at $1kg/cm^2$ which is $1.034 \times 10^5 Pa$, and sterilize about 20 min. After sterilization, cut off the power. When the pressure drops to 0, the lid should immediately be open and the items are removed.

2. Filter sterilization

Some growth regulators like amino acids and vitamins are heat labile and can be destroyed in autoclaving with the rest of the nutrient medium. Therefore, it is sterilized by filtration through a sieve or a filtration assembly using filter membranes of 0.22 μm to 0.45 μm size.

3. Irradiation

It can only be carried out under the condition where UV radiation is available. Consequently, its use is restricted generally to purchased consumables like petri dishes and pipettes. UV lights may be used to kill organisms in rooms or areas of work benches in which manipulation of cultures is carried out. It is, however, dangerous and should not be turned on while any other work is in progress. UV light of some wavelengths can damage eyes and skin.

4. Laminar airflow cabinet

This is the primary equipment used for aseptic manipulation. This cabinet should be used for horizontal air-flow from the back to the front. Air is drawn in electric fans and passed through the coarse filter and then through the fine bacterial filter (HEPA). High Efficiency Particulate Air Filter is an apparatus designed such that the air-flow through the working place flows in direct lines. Care is taken not to disturb this flow too much by vigorous movements. Before commencing any experiment it is desirable to clean the working surface with 70% alcohol. The air filters should be cleaned and changed periodically.

V. Experimental Results

Taking certain amount of medium to culture at 37 ℃ for 24 h. If there is no bacterial growth, the medium can be stored for standby. Slant medium from the pot is put on the slope until it solidifies, then place in 37 ℃ for 24 h. If there is no bacterial growth , it can be saved for backup.

Questions:

(1) Why can not items requiring sterilization be placed too close in sterilizer?

(Answer is omitted.)

(2) What is the sterilization temperature? What should be noticed after sterilization?

(Answer is omitted.)

实验十七　从大肠杆菌中小规模制备质粒

一、实验目的

学习从大肠杆菌小规模制备质粒的方法。

二、实验原理

质粒是存在于许多微生物中的染色体外的双链闭合环状 DNA。其通常在细胞中有不同寻常的特性，如结合能力，抗生素耐药性转移，异型化学物质的降解，中和毒素物质等。在自然界中，质粒分子的大小为 2~100 Kb。质粒有自己的复制子，并能利用宿主 DNA 聚合酶进行繁殖。然而，分子克隆质粒与天然质粒是不同的，因为分子克隆质粒小并有多个限制性酶切位点。在克隆实验中，质粒 DNA 需要提取（从细菌宿主，主要是大肠杆菌）。碱性裂解法是小规模制备高纯度细菌质粒的方法。这个方法常用于从细菌细胞悬浮液中提取质粒 DNA。质粒是染色体外的相对较小的超螺旋 DNA 分子，然而细菌染色体 DNA 比较大，超螺旋也较少。因此不同的拓扑结构可以允许选择性沉淀染色体 DNA、细胞蛋白质和 RNA 分子。在碱性条件下，核酸和蛋白分离，在加入醋酸钾中和后它们可以复性。染色体由于太大而不能正确地复性而除去。因此质粒被有效地从染色体中提取出。

十二烷基硫酸钠（SDS）和氢氧化钠可以完全裂解细胞。葡萄糖作为 pH 缓冲剂来调整 pH 值。具有高分子量的染色体 DNA 可以有选择地变性。醋酸钠用于中和碱液，这样大量的染色体 DNA 复性后形成一个不溶性颗粒。同时，高浓度的醋酸钠也可导致蛋白质十二烷基磺酸钠复合物和高分子量的 RNA 沉淀。这样三个主要不溶物——染色体 DNA、蛋白质与十二烷基磺酸钠复合物和高分子量 RNA，可以通过微型离心机离心分离。为了从上层清液中获得质粒 DNA，可以用乙醇进行沉淀。这种小通量提取方法可提取 5~10 μg 质粒 DNA。这个方法也能通过扩大来提取更大量的 DNA（或 RNA）。提取的结果用凝胶电泳分析进行验证。

三、实验材料

培养的隔夜生长细菌、无菌离心管、无菌枪头、微量吸液管、溶液Ⅰ、溶液Ⅱ、溶液Ⅲ、核糖核酸酶、酚、氯仿、异戊醇、异丙醇、醋酸钠、乙醇和 TE 缓冲液。

四、实验步骤

（1）养 2 mL 菌液并加适量的抗生素，37℃震荡培养 4~5 h，直到生长对数期（检查培养液的浊度）。

（2）往离心管（1.5 mL）放入 1.5 mL 培养液，10 000 转速离心 2 min，去除上清液；继续加入其余的 3 mL 培养液于同一离心管，每次 1.5 mL。

（3）用 100 μL 的溶液Ⅰ（Tris、EDTA、葡萄糖）悬浮细胞（可用涡旋器悬浮）。

（4）立即加入 200 μL 刚做好的溶液Ⅱ（0.4 mol/L 氢氧化钠和 2% 十二烷基磺酸钠，1∶1）。颠倒管混匀。

（5）再加 150 μL 的溶液Ⅲ（冰冷）到每个管，通过翻转混合，以 12 000 r/min 离心 15 min。转移上层清液到新管（小心以避免触动分界线），加 5 μL 的核糖核酸酶（10 mg/mL）到每个管，通过翻转混合，在 37℃下培养（水浴）1 h。

（6）加入等体积的氯仿和异戊醇（200 μL），用涡旋器混合。以 12 000 r/min 离心 15 min。将新鲜上清液小心加入离心管中，再加入等体积（400 μL）异丙醇，然后加入 0.1 体积（40 μL）醋酸钠（pH5.2），翻转混合。保持在 −20℃（过夜）。

（7）以 12 000 r/min，在 4℃下，离心 15 min。倒出上层清液，添加 200 μL 冰冷的 70% 乙醇，混合翻转，以 12 000 r/min，在 4℃下，离心 5 min。吸出上层清液，沉淀干燥。

（8）将沉淀溶解在 40 μL TE（10 mM 三异丙基乙磺酰 +1 mM 乙二胺四乙酸）里，通过轻摇混合。在 −20℃下储存。

注意事项：

（1）添加溶液Ⅱ后不能再涡旋试管。

（2）苯酚必须为饱和酚，其 pH 值达到 8.0 且质量要很好。

（3）苯酚和氯仿处理后，在去除上层水相时不要破坏其白色界面。

（4）当用70%的乙醇进行洗涤时，不能打破DNA颗粒，这一步主要是洗去DNA颗粒残留的乙醇和盐类。

（5）由于大量的核糖核酸污染，在添加苯酚氯仿之前，可以添加核糖核酸酶以降解RNA。

（6）在随后的操作中，可随着限内切酶一同添加核糖核酸酶。

（7）有时质粒在TE中变得难以溶解，这时把它放入冰箱中，第二天，DNA就容易溶解了。

五、实验结果

进行凝胶电泳分析来验证实验结果。

思考题：

（1）在以碱法提取质粒的过程中，EDTA、溶菌酶、NaOH、SDS、乙酸钾、氯仿等试剂的作用是什么？

（答案略。）

（2）在提取质粒的过程中，应注意哪些操作？为什么？

（答案略。）

附录：

试剂：

溶液 I :（100 mL）

项　目	相对分子质量	针对 100 mL / g
Tris (25mM)	121.10	0.303
EDTA (10mM)	372.00	0.372
Glucose (50mM)	180.16	0.901

称量以上盐并溶于80 mL的dd水，使用1 mol/L盐酸调整pH值为8.0。加入dd水使体积达到100 mL。高压灭菌，然后室温储存。

溶液 II :（每次现做）

NaOH 0.2 M

SDS 1.0%

准备0.4 M氢氧化钠并存储在一个塑料试剂瓶中。准备2%十二烷基磺酸钠并进行高压蒸汽灭菌，在使用前将它们以 1∶1 的比率混合。氢氧化钠不能高压灭菌。

溶液Ⅲ：[3M 乙酸钾（pH=5.5）]

称量 29.4 g 乙酸钾并溶解在 25~30 mL 的蒸馏水中。以冰醋酸调整 pH 值，并使体积达到 100 mL。高压灭菌，然后 4℃储存。

核糖核酸酶

溶解胰核糖核酸酶（核糖核酸酶 A），使浓度为 10 mg/mL（10 mM Tris，pH7.5，15 mM NaCl），加热到 100℃并保持 15 min（变性脱氧核糖核酸酶）。慢慢冷却到室温。均等分配，在 −20℃下存储。

氯仿∶异戊醇：按氯仿∶异戊醇 =24∶1 的比率制备。

3M 醋酸钠（pH=5.2）100 mL：称量 24.61 g 醋酸钠并溶于 80 mL 的重蒸馏水中。用冰醋酸调整 pH 值。使体积到 100 mL。高压灭菌，然后 4℃储存。

TE（0.1X）pH 8.0 100 mL：

Tris HCl（1mM）——从 1M 储液（pH8.0）中取 100 μL。

EDTA（0.1M）——从 0.5M 储液（pH8.0）中取 20 μL。

加无菌双蒸馏水 98.8 mL。

Experiment 17   Small Scale Plasmid Preparation Through E. Coli

Ⅰ. Experimental Objectives

To study the method of small scale plasmid preparation from E. coli.

Ⅱ. Experimental Principles

Plasmids are extrachromosomal, double-stranded, closed-circular DNA presented in many microorganisms. Plasmids are usually present in the cell conferring extraordinary properties, like the ability of conjugation, conferring antibiotic resistance, degradation of xenobiotic substances, and production of substances to neutralize toxins, etc. In nature, plasmids are large molecules with sizes ranging from 2 Kb to 100 Kb. Plasmids have origin of replication and multiply by utilizing the long-leaved enzymes of the host DNA polymerizes. However, the plasmids used in molecular cloning are different from the natural plasmids, as they are small in size and offer multiple restriction sites. Plasmid DNA needs to be extracted (from bacterial hosts, mostly E. coli) almost routinely in cloning experiments. Alkaline lysis plasmid miniprep is a procedure used to prepare bacterial plasmids in a highly purified form. This method is used to extract plasmid DNA from bacterial cell suspensions. Plasmids are relatively small extra chromosomal supercoiled DNA molecules while bacterial chromosomal DNA is much larger and less supercoiled. Therefore, the difference in topology allows for selective precipitation of the chromosomal DNA, cellular proteins from plasmids and also RNA molecules. Under alkaline conditions, both nucleic acids and proteins denature. They are renatured when the solution is neutralized by the addition of potassium acetate. Chromosomal DNA is precipitated out because the structure is too big to renature correctly; hence plasmid DNA is extracted efficiently in the solution.

The cells are then lysed completely with sodium dodecyl sulfate (SDS) and NaOH. Glucose is also used as a pH buffer to control the pH. Chromosomal DNA, which remained in a high molecular weight form, is selectively denatured. Acid sodium acetate is used to neutralize the lysate as the mass of chromosomal DNA features and coagulates to form an insoluble pellet. At the same time, high concentration of sodium acetate also results in the precipitation of protein-SDS complexes and high molecular weight RNA. By now, three major contaminants−chromosomal DNA, protein-SDS complexes and high molecular weight RNA can be removed by spinning in a microcentrifuge. In order to recover plasmid DNA in the supernatant, ethanol precipitation is carried out. A mini prep usually yields 5−10 μg. This can be scaled up to a midi prep or a maxi prep, which will yield much larger amounts of DNA (or RNA). A gel electrophoresis analysis is conducted to verify the results.

Ⅲ. Experimental Materials

Overnight grown bacterial culture; sterile eppendorf tubes; sterile microtips; and micropipette; Solution Ⅰ, Ⅱ and Ⅲ; RNAse, phenol, chloroform, isoamyl alcohol, isopropanol, sodium acetate, ethanol, and TE buffer.

Ⅳ. Experimental Procedures

(1) Grow 2 mL culture with appropriate antibiotics for 4−5 h at 37℃ in a shaker till log phase (Check for the turbidity of the culture).

(2) Pipette 1.5 mL culture into an eppendorf tube (1.5 mL), spin at 10,000 for 2 min, and remove the supernatant, spin down the rest of 3 mL culture in the same eppendorf tube, 1.5 mL at a time.

(3) Resuspend the cells in 100 μL of Solution I (Tris, EDTA, and Glucose) (or Suspend by vigorous vortexing).

(4) Immediately add 200 μL of freshly prepared Solution II (0.4 mol/L NaOH and 2% SDS, 1 : 1). Mix by inverting the tube.

(5) Add 150 μL of Solution Ⅲ (ice cold) to each tube, mix by inverting, spin at 12,000 rpm for 15 min. Transfer the supernatant to a fresh tube (carefully by avoiding the interphase), add 5 μL of RNAse (10 mg/mL) to each individual tube, mix by inverting, incubate at 37℃ (water bath) for 1 h.

(6) Add equal volume of chloroform and isoamyl alcohol (200 μL), mix by vortexing vigorously, spin at 12,000 rpm for 15 min. Transfer the supernatant carefully to a fresh eppendorf tube, add equal volume (400 μL) of isopropanol and then add 0.1 volume (40 μL) of sodium acetate (pH 5.2) to each tube, mix by inverting, and keep it in −20℃ (Over night).

(7) Spin at 12,000 rpm, 4℃ for 15 min. Discard the supernatant; add 200 μL of ice cold 70% ethanol, mix by inverting, spin at 12,000 rpm and 4℃ for 5 min. Discard the supernatant by using pipette, and then dry the pellet.

(8) Dissolve the pellet in 40 μL TE (10 mM Tris+1 mM EDTA), mix them by tapping and give a short spin. Store the solution at −20℃ .

Attention:

(1) Do not vortex the tubes after the addition of solution Ⅱ.

(2) The phenol has to be tris-saturated to pH 8.0 and has very good quality.

(3) Do not disturb the whitish interface while removing the upper aqueous phase after phenol and chloroform treatments.

(4) While washing with 70% ethanol, do not break the DNA pellet. This is for washing the pellet only to remove traces of ethanol and salts.

(5) This preparation will contain a lot of RNA contamination. DNAse-free RNAse may be added before the phenol: chloroform step to digest the RNA.

(6) During subsequent manipulations RNAse may be added, along with the restriction enzyme.

(7) Sometimes it may become difficult to dissolve the plasmid preparation in TE. Keep it in the freezer overnight. In the next day, the DNA will easily dissolve.

Ⅴ. Experimental Results

A gel electrophoresis analysis is conducted to verify the results.

Questions:

(1) What is the function of EDTA, lysozyme, NaOH, SDS, potassium acetate,

phenol/chloroform and other agents in the process of extraction of plasmid by alkali method?

(Answer is omitted.)

(2) What should be paid attention to in the process of plasmid extraction? Why?

(Answer is omitted.)

Appendix:

REAGENTS

Solution I: (100 mL)

Item	Mol. Wt	For 100 mL / g
Tris (25 mM)	121.10	0.303
EDTA (10 mM)	372.00	0.372
Glucose (50 mM)	180.16	0.901

Weigh the above salt and dissolve it in 80 mL of dd water and adjust it to pH 8.0 using 1 mol/L HCl. Make up the volume to 100 mL. Autoclave and store it at the room temperature.

Solution II : (prepare fresh each time)

NaOH 0.2 M

SDS 1.0%

Prepare 0.4 M NaOH and store it in a plastic reagent bottle. Prepare 2% SDS and autoclave. Mix them in 1 : 1 ratio before using. Do not autoclave NaOH.

Solution III : [3 M potassium acetate (pH 5.5)]

Weigh 29.4 g of potassium acetate and dissolve in 25mL to 30mL distilled water. Adjust the pH with glacial acetic acid and make up the volume to 100 mL. Autoclave and store them at 4℃.

RNAse

Dissolve pancreatic RNAse (RNAse A) to a concentration of 10 mg/mL (10 mM Tris, pH 7.5, 15mM NaCl), heat to 100℃ and hold for 15 min in a boiling water bath (to denature DNASE). Allow it to cool slowly to the room temperature. Dispense it into aliquots and store it at −20℃ .

Chloroform: Isoamyl alcohol: Prepare Chloroform : Isoamyl alcohol in 24 : 1 ratio.

3M sodium acetate (pH 5.2) 100 mL: Weigh 24.61 g of sodium acetate and dissolve it in 80 mL of double distilled water. Adjust the pH with glacial acetic acid. Make up the volume to 100 mL. Autoclave it and store it at 4°C .

TE (0.1X) pH 8.0 100 mL:

Tris HCl (1 mM)—100 µL from 1 M stock (pH 8.0)

EDTA (0.1 M) —20 µL from 0.5 M stock (pH 8.0)

Add 98.8 mL of sterile double distilled water.

实验十八　MS培养基的准备

一、实验目的

掌握配置培养基各种母液和植物生长素的方法。

二、实验原理

基础培养基提供了植物生长和发育所需要的化合物，其中某些化合物可以通过完整的植物合成，但不能单独由植物组织合成。组织培养基的构成包括95%水、大量和微量营养素、维生素、氨基酸和糖等成分。植物细胞可利用培养基中的营养成分来合成其所需的有机分子，或在酶反应中用作酶促剂等。植物所需大量营养素的量在毫摩尔（mm）每升，而所需微量营养素浓度则低得多（微摩尔每升，μm/L）。维生素是有机物质，是作为关键代谢的功能的酶的组成部分。糖几乎是所有组培作物中胚的生长发育必要元素，在组培中因各种各样的原因其不能通过光合作用合成。MS培养基是最合适、最常用的植株再生基本组织培养基。在固体或液体培养基中，植物生长调节剂以非常低的浓度（0.1~100 μm/L）来调控外植体的芽和根的形成和发育。生长素和细胞分裂素是组织培养过程中最重要的两类植物生长调节剂。生长素和细胞分裂素相对效应比决定了培养组织的形态发生。

三、实验材料

琥珀瓶、塑料烧杯（100 mL，500 mL和1 000 mL）、量筒（500 mL）、玻璃烧杯（50 mL）、一次性注射器（5 mL）、一次性注射器过滤器（0.22 μm）、Eppendorf管（2 mL）、萘乙酸。

四、实验步骤

营养盐和维生素都准备成储液（20× 或 200× 浓度）。储液存放在 4℃条件。

1. MS 大量母液（20×）的配置

项目	大量元素	每升培养基所含的量 /mg	500 mL 母液 (20×) 所含量 /g
1	NH_4NO_3	1 650	16.5
2	KNO_3	1 900	19.0
3	$CaCl_2 \cdot 2H_2O$	440	4.4
4	$MgSO_4 \cdot 7H_2O$	370	3.7
5	KH_2PO_4	170	1.7

2. MS 微量母液（200×）的配置

项目	微量元素	每升培养基所含的量 /mg	500 mL 母液 (200×) 所含量 /mg
1	H_3BO_3	6.200	620.0
2	$MnSO_4 \cdot 4H_2O$	22.300	2 230.0
3	$ZnSO_4 \cdot 4H_2O$	8.600	860.0
4	KI	0.830	83.0
5	$Na_2MoO_4 \cdot 2H_2O$	0.250	25.0
6	$CoCl_2 \cdot 6H_2O$	0.025	2.5
7	$CuSO_4 \cdot 5H_2O$	0.025	2.5

3. MS 维生素母液（200×）的配置

项目	MS 维生素	每升培养基所含的量 /mg	500 mL 母液 (200×) 所含量 /mg
1	硫胺素	0.1	10
2	烟酸	0.5	50
3	甘氨酸	2.0	200
4	维生素 B6	0.5	50
5	肌醇	100.0	10 000

4. 500 mL 铁盐母液 (200×) 的配置

溶解 3.725 mg 乙二胺四乙酸二钠（乙二胺四醋酸二钠）在 250 mL 蒸馏水。将 2.785 mg 七水硫酸亚铁溶解在 250 mL 蒸馏水中。将乙二胺四乙酸二钠溶液煮沸，最后加上硫酸亚铁溶液，轻轻搅拌。

5. 植物生长调节剂储藏液

热不稳定的植物生长调节剂通过细菌过滤器膜过滤（0.22 μm）除菌，并添加到蒸压灭菌冷却后的培养基中（低于 60℃）。上面所列的植物生长调节剂按表中方法溶解，然后经过滤器过滤（0.22 μm），并存储在 −20℃。

五、实验结果

（1）装瓶：将配制好的母液分别装入试剂瓶中，贴好标签，注明各培养基母液的名称、浓缩倍数、日期。注意将易分解、氧化的溶液，放入棕色瓶中保存。

（2）贮藏：4℃冰箱。

思考题：

配置 MS 培养基的铁盐母液为什么要加热且不停搅动？

（答案略。）

Experiment 18   Preparation of MS Medium

I . Experimental Objectives

Master the methods of prefaring all kinds of solutions of MS basal and auxin configuration medium.

II . Experimental Principles

The basal medium is formulated so that it provides all of the compounds needed for plant growth and development, including certain compounds that can be made by an intact plant, but not by an isolated piece of plant tissue. The tissue culture medium consists of 95% water, macro- and micro-nutrients, vitamins, amino acids, sugars and etc. The nutrients in the media are used by the plant cells as building blocks for the synthesis of organic molecules, or as catalyzators in enzymatic reactions. The macronutrients are required in millimolar (mm/L) quantities while micronutrients are needed in much lower (micromolar, μm/L) concentrations. Vitamins are organic substances that are parts of enzymes or cofactors for essential metabolic functions. Sugar is essential in vitro growth and development as most plant cultures are unable to photosynthesize effectively for a variety of reasons. Medium (MS) is the most suitable and commonly used basic tissue culture medium for plant regeneration. Plant growth regulators (PGRs) at a very low concentration (0.1 to 100 μm/L) regulate the initiation and development of shoots and roots on explants on semisolid or in liquid medium cultures. The auxins and cytokinins are the two most important classes of PGRs used in tissue culture. The relative effects of auxin and cytokinin ratio determine the morphogenesis of cultured tissues.

Ⅲ. Experimental Materials

Amber bottles, plastic beakers (100 mL, 500 mL and 1,000 mL), measuring cylinders (500 mL) , glass beakers (50 mL), disposable syringes (5 mL), disposable syringe filter (0.22 μ m), autoclaved Eppendorf tubes (2 mL), and naphthalene acetic acid.

Ⅳ. Experimental Procedures

Nutrient salts and vitamins are prepared as stock solutions (20 × or 200 × concentration) as specified. The stocks are stored at 4℃ .

1. MS major salts STOCK (20 ×)

Item	MS major salt	Content per litre medium /mg	Content in 500 mL stock (20 ×) /g
1	NH_4NO_3	1,650	16.5
2	KNO_3	1,900	19.0
3	$CaCl_2 \cdot 2H_2O$	440	4.4
4	$MgSO_4 \cdot 7H_2O$	370	3.7
5	KH_2PO_4	170	1.7

2. MS minor salts STOCK (200 ×)

Item	MS minor salt	Content per litre medium /mg	Content in 500 mL stock (200 ×) /mg
1	H_3BO_3	6.200	620.0
2	$MnSO_4 \cdot 4H_2O$	22.300	2 230.0
3	$ZnSO_4 \cdot 4H_2O$	8.600	860.0
4	KI	0.830	83.0
5	$Na_2MoO_4 \cdot 2H_2O$	0.250	25.0
6	$CoCl_2 \cdot 6H_2O$	0.025	2.5
7	$CuSO_4 \cdot 5H_2O$	0.025	2.5

3. MS Vitamins STOCK (200×)

Item	MS Vitamin	Content per litre medium /mg	Content in 500 mL stock (200×)/mg
1	Thiamine	0.1	10
2	Nicotinic acid	0.5	50
3	Glycine	2.0	200
4	Vitamin B6	0.5	50
5	iNositol	100.0	10 000

4. Iron, 500mL STOCK (200×)

Dissolve 3.725 mg of Na_2 EDTA (Ethylenediaminetetra acetic acid, disodium salt) in 250 mL dH_2O. Dissolve 2.785 mg of $FeSO_4 \cdot 7H_2O$ in 250 mL dH_2O. Boil Na_2 EDTA solution and add $FeSO_4$ solution into it and stir gently.

5. Plant growth regulator STOCK

The heat-labile plant growth regulators are filtered through a bacteria-proof membrane (0.22 μm) filter and added to the autoclaved medium which has cooled enough (less than 60 ℃). The desired amount of plant growth regulators is dissolved through the above methods. The solutions are passed through disposable syringe filter (0.22 μm) and then stored at −20℃ .

V . Experimental Results

(1) Put the stock mother liquor into the reagent bottles: Label and specify the name, the concentration and date of the liquor. Note that the solution which is easy to decompose and oxidate, is stored in brown bottles.

(2) Condition of storage: 4℃ in refrigerator.

Question:

Why does salt solution configuration of MS culture medium need to be kept being heated and stirred?

(Answer is omitted.)

第四部分

实验十九　脱氧核糖核酸 (DNA) 的鉴定
——孚尔根（Feulgen）反应

一、实验目的

学习和掌握孚尔根反应染色法，鉴定植物细胞核内 DNA 的存在。

二、实验原理

染色体是遗传物质的载体，其主要化学成分是脱氧核糖核酸（DNA）。孚尔根染色是罗伯特孚尔根发明的染色体染色技术，在组织学用于鉴定细胞标本的染色体或 DNA。它取决于 DNA 的酸水解，因此在固定剂的使用上应避免强酸。该技术关键在样本材料的稀盐酸处理，水解 DNA 和脱下碱基。留下的糖结构反应为醛，接着醛基与希夫试剂缩合，反应成典型的红色。

三、实验材料

洋葱或大蒜的根尖、复式显微镜、载玻片、盖玻片、镊子、滤纸、刀片、45% 乙酸、1-M HCl、卡诺氏固定液 [乙醇混合物（冰醋酸为 3∶1 或 9∶1）]、希夫氏试剂（无色碱性品红液）和漂洗液。

四、实验步骤

1. 取材与固定

待大蒜根尖长 1 cm 时，剪下根尖，投入卡诺固定液中固定 2~24 h。固定后，保存在 70% 的酒精中，放入冰箱中冷藏供用。

2. 水解

试管 1：取大蒜根尖数条，投入试管，先用清水洗 3 次，常温下换 1 mol/L HCl 洗一次 1 min。加入预热 60℃ 的 1 mol/L HCl 浸没根尖，放入恒温水浴锅中在 60℃ 下水解 10 min（水解时间视材料而定）。然后在常温下的 1 mol/L HCl 洗一次，再用清水将根尖洗三次。

试管 2（对照）：取大蒜根尖数条，投入试管，先用清水洗 3 次，入预热 60℃ 的蒸馏水浸没根尖，放入恒温水浴锅中在 60℃ 下水解 10 min。以下各步骤两试管相同。

3. 染色

两试管中均滴加希夫试剂，浸没根尖，立即置于暗处处理 0.5~1 h，然后用漂洗液换洗 3 次，每次 2 min；再用蒸馏水换洗 2 次，每次 2 min。

4. 镜检

分别取上述两试管中的根尖，置于载玻片上，滴加 1 滴 45% 醋酸，然后压片。把载玻片放在显微镜观察台上，观察。

注意事项：

注意水解时间和温度。

五、实验结果

画出经孚尔根染色的细胞图像，比较根尖经盐酸水解处理与否所获得的结果。

附录：

Schiff 试剂的配制

称 1 g 碱性品红于 200 mL 煮沸的蒸馏水中，5 min 后断电使其冷却至 55℃ ~50℃，过滤到一个棕色的试剂瓶中，加入 1 mol/L HCl 20 mL，继续冷却至 25℃，加入 3 g 偏重亚硫酸钠，摇动瓶子使其溶解。密闭瓶口，置冰箱内（4℃左右），18~24 h 后检查，试剂如果是透明无色或呈浅黄色，则可使用。如有不同程度的红色未褪，可加入 1 g 活性炭，强烈震荡一分钟，仍在低温下静置过夜，然后用滤纸过滤后使用。密封瓶口，包以黑纸，在5℃以下冰箱内可以保存半年。

　　漂洗液

　　在 200 mL 蒸馏水中，加入 10 mL 1 mol/L HCl 和 10 mL 10% 亚硫酸钠溶液。此液在临用前配制。

Experiment 19   Identification of Deoxyribonucleic Acid (DNA)—Feulgen Reaction

I . Experimental Objectives

Learn and master the Feulgen reaction staining method and identification of plant nuclei in the presence of the DNA chromosome.

II . Experimental Principles

Chromosome is the carrier of genetic materials; its main chemical composition is deoxyribonucleic acid (DNA). Feulgen stain is a staining technique invented by Robert Feulgen and used in histology to identify chromosomal materials or DNA in cell specimens. It depends on acid hydrolysis of DNA; therefore using strong acids as fixation agents should be avoided. The technique involves treating sections with diluting hydrochloric acid, hydrolyzing DNA and removing the bases. The sugar remains and reacts as an aldehyde, including condensation with Schiff's reagent when it is subsequently applied. The typical red coloration where DNA is present, is thus formed.

III . Experimental Materials

Root of onions or garlic, compound microscope, slides and cover slips, forceps, filter paper, scalpel, 45% acetic acid, 1-M HCl (Hydrochloric acid), Carnoy's fixative(Mixture of ethyl alcohol:glacial acetic acid in proportions of 3 : 1 and 9 : 1), Schiff reagent, sulphuric acid.

Ⅳ. Experimental Procedures

1. Sampling fixation

When the garlic root tip grows to the length of 1 cm, cut the root tip and put it into Carnoys fixative for 2-24 h. After fixation, preserve it in 70% alcohol and store it in the freezer for later use.

2. Hydrolysis

Tube 1: Put several garlic root tips into tube1 and rinse 3 times with water. Rinse them again with 1 mol/L-HCl at room temperature for 1 min. Place the samples in preheated 1 mol/L-HCl at 60°C for 10 min (this hydrolysis time will vary and depend on the samples). Then rinse them in 1 mol/L-HCl at room temperature for 1 min and lastly rinse in water three times.

Tube 2(Control): Put several garlic root tips into tube 2 and rinse them 3 times with water. Place the samples in preheated water at 60°C for 10 min. The following steps are the same for the above two tubes.

3. Staining

Both tubes were dripped with Schiff reagent which immersed the sample, and immediately placed the tubes in the dark for 0.5-1 h and then rinse with Sulphurous acid three times for 2 min each. Then rinse with distilled water two times for 2 min.

4. Observing under microscope

Take the root tips from the above two tubes, place them on a slide, add 1 drop of 45% acetic acid, and then put a lid on the slide. Place the slide on the stage of a microscope and observe.

Attention:

Note the time and temperature of hydrolysis.

Ⅴ. Experimental Results

Draw the images of Feulgen stained cells and compare the control cells with those stained by the Feulgen reaction.

Appendix:

Preparation of Schiff reagent

Take 1 g basic fuchsin, and add it into 200 mL of boiling distilled water. About 5 minutes later plug it off and leave it cooling to 55℃ −50℃ , then filter it into a brown bottle and add 20 mL 1 mol/L HCl, continue to cool to 25℃ . Add 3 g sodium pyrosulfite and shake to dissolve. At last seal and store it at the refrigerator (4 ℃). After 18 to 24 h, if the reagent is a transparent and colorless or light yellow, it can be used. If there is a certain degree of unfaded red, add 1 g activated carbon and strongly shake for a minute. Store at the low temperature overnight, and use after filter. It can be saved for half a year in 5 ℃ refrigerator if sealed and wrapped with black paper.

Sulphurous acid

Add 10 mL 1 mol/L HCl and 10 mL 10% sodium Sulphurous acid into the 200 mL distilled water. Prepare the liquid before using.

实验二十　植物基因组 DNA 的提取

一、实验目的

掌握从高等植物细胞中制备基因组 DNA 的基本原理，熟悉从高等植物中提取基因组 DNA 的技术流程。

二、实验原理

本实验介绍的方法是 CTAB（cetyltriethyl ammonium bromide）法。CTAB 是一种去污剂，能裂解植物细胞，并能跟核酸在低盐溶液中形成不可溶的复合物。而在低盐条件下，多糖、酚类化合物和其他杂质留在上清液中被洗去。CTAB-核酸溶解于高盐溶液中，并在乙醇或异丙醇中沉淀下来。本实验主要分为 3 步：细胞膜裂解；DNA 提取；DNA 沉淀。

三、实验材料

幼嫩的幼苗、水浴锅、液氮罐、离心机、移液枪、电泳设备、核酸紫外检测仪等、研钵、杵子、1.5 mL 离心管、移液枪枪头、玻璃棒等。

液氮，核糖核酸酶（RNase），2×CTAB 抽提缓冲液（2%CTAB，1.4 mol/L NaCl，100 mmol/L Tris-Cl，pH8.0，20 mmol/L EDTA，2% 巯基乙醇），氯仿:异戊醇（24:1），异丙醇，无水乙醇，70% 乙醇，TE（10 mmol/L Tris·Cl，1 mmol/L EDTA，pH8.0）缓冲液。

四、实验方法

（1）将 0.2 g 冻存的叶片放于研钵中，用液氮速冻并研磨成粉末。

（2）加入 500 μL 左右 CTAB 缓冲液，转入到 1.5 mL 离心管中。

（3）于 65℃水浴保温 30 min 以上，并间或轻摇混匀。

（4）加入等体积的氯仿∶异戊醇（24∶1），轻轻混匀并放在常温下试管倒相器上约 20 min。

（5）以速度 12 000 rpm 离心 10~20 min。

（6）将上清液移入另一干净的离心管中，加入 300 μL 的异丙醇，小心摇动离心管，置于冰上 10 min。

（7）以速度 12 000 rpm 离心 5 min，收集 DNA 沉淀。

（8）倒去上清液，用吸水纸将试管内壁擦干。

（9）干燥后的 DNA 沉淀溶于含 20 μg/mL RNase A 的 500 μL TE 缓冲液中，于 37℃消化 RNA 1 h。

（10）加入 1/10 体积的 3 mol/L NaAc（pH5.2）溶液和 2 倍体积预冷的无水乙醇，小心混匀，置于冰上 5 min。

（11）以速度 12 000 rpm 离心 5 min，收集 DNA 沉淀。

（12）弃上清液，用吸水纸将试管内壁擦干。

（13）干燥后溶于 100 μLTE 缓冲液中，可在 −20℃保存备用。

五、实验结果

描述 DNA 提取过程每一步的现象。

取小量 DNA 溶液，在 0.8% 的琼脂糖凝胶上电泳分离，并在核酸紫外检测仪上观察。

思考题：

（1）巯基乙醇的作用是什么？可以用什么来替代吗？

（答案略。）

（2）Tris-HCl（pH8.0）和 NaCl 的作用是什么？

（答案略。）

（3）乙醇的作用是什么？

（答案略。）

Experiment 20　Extraction of Plant Genome DNA

Ⅰ. Experimental Objectives

Master the basic principles of preparing genomic DNA and familiarize with the technical process of extracting genomic DNA from higher plants.

Ⅱ. Experimental Principles

This experiment is introduced by Methods of CTAB (cetyltriethyl ammonium bromide). Plant cells can be lysed with the ionic detergent cetyltrimethyl ammonium bromide (CTAB), which forms an insoluble complex with nucleic acids in a low-salt environment. Under these conditions, polysaccharides, phenolic compounds and other contaminants remain in the supernatant and can be washed away. The DNA complex is solubilized by raising the salt concentration and precipitated with ethanol or isopropanol. In this section, the principles of these three main steps—lysis of the cell membrane, extraction of the genomic DNA and its precipitation will be described.

Ⅲ. Experimental Materials

Plant seedlings, water bath, the liquid nitrogen tank, centrifuge, pipette, electrophoresis equipment, nucleic acid ultraviolet detector, mortar, pestle, 1.5mL centrifuge tube, pipetting gun head, and glass rod etc.

Liquid nitrogen, RNase, CTAB buffer (2%CTAB, 1.4 mol/L NaCl, 100 mmol/L Tris-Cl, pH8.0, 20 mmol/L EDTA, 2% 2-Mercaptoethanol), Chloroform: isoamyl alcohol (24:1), isopropanol, 100% ethanol, 70% ethanol, TE(10mmol/L Tris · Cl, 1 mmol/L EDTA, pH8.0).

IV . Experimental Procedures

(1) Grind 0.2 g of frozen leaves to a very fine powder with liquid nitrogen using mortar and pestle.

(2) Add about 500 μL of CTAB buffer and transfer to a 1.5 mL tube.

(3) Incubate at 65°C for 30 min with occasional vigorous shaking.

(4) Add 500 μL of Chloroform: isoamyl alcohol (24 : 1), shake well, and place on a tube inverter at room temperature for about 20 min.

(5) Centrifuge at 12,000 rpm for 10-20 min.

(6) Transfer the aqueous phase to a fresh tube, add 300 μL of isopropanol, mix, and place on ice for 10 min.

(7) Centrifuge at 12,000 rpm for 5 min to collect the precipitate.

(8) Discard the supernatant and dry the inside of the tube with a paper towel.

(9) Add 500 μL of TE buffer (Contained 20 μg/mL RNase A) and dissolve the precipitate under 37 °C for 1 h.

(10) Add 1/10 volume of 3 mol/L NaAc (pH5.2) and 2 volumes of 100% Ethanol to each tube and incubate on ice for 5 min.

(11) Centrifuge at 12,000 rpm for 5 min to collect the precipitate.

(12) Discard the supernatant and dry the inside of the tube with a paper towel.

(13) Add 100 μL of TE Buffer and dissolve DNA. Save for backup in −20 °C .

V . Experimental Results

Description of the phenomenon in process of DNA extraction at every step.

Take a small amount of DNA solution, put it in 0.8% agarose gel by electrophoresis, and observe the concentration and purity of DNA under UV spectroscopy.

Questions:

(1) What is the effect of mercaptoethanol? What can be used to replace it?

(Answer is omitted.)

(2) What is the role of Tris-HCl (pH8.0) and NaCl?

(Answer is omitted.)

(3) What is the role of alcohol?

(Answer is omitted.)

实验二十一　利用紫外光谱测定 DNA 的浓度和纯度

一、实验目的

通过本实验掌握利用紫外光谱测定 DNA 的浓度和纯度的方法。

二、实验原理

在分子操作过程中对 DNA 样品的浓度及相对纯度进行精确定量非常重要，无论是将纯化的 DNA 片段连接到克隆载体中，还是对小量制备的质粒或 PCR 扩增得到的 DNA 进行测序或进行其他更专业的操作，DNA 样品的浓度及相对纯度在很大程度上会影响到实验结果的好坏及实验结果的重现。UV 分光光度法是一种精确、快速并且无破坏性的测定 DNA 浓度的方法，它所测得的 DNA 的最低浓度可达 2.5 μg/mL，因此在分子生物学实验中经常采用。

DNA 分子中碱基的紫外吸收图谱在 260 nm 处有一个特征吸收峰。这种吸收和 DNA（或 RNA）量成正比。应用 1cm 光路，DNA 在 260 nm 波长处的消光系数为 20。在 1 cm 石英比色皿中，其 50 μg/mL 双链 DNA 溶液在 260 nm 处吸收值为 1。DNA 双螺旋和单链分子之间的相互转换会使它的吸收水平产生一定的变化，但是这种偏差可以用一个特定的公式校正。这种方法方便且相对准确。由于色氨酸残基的吸收，蛋白质的最大吸收大约在 280 nm 处。因此，A260/A280 的比值可用来衡量 DNA 制剂的纯度，并且比值应在 1.65~1.85。较低的比值说明了 DNA 制剂被蛋白质污染。由于苯酚在 270 nm 处有最大吸收，如果 DNA 制剂被苯酚污染，则 A260 将会异常高，会导致 DNA 浓度被高估。

三、实验材料

DNA 样品，彼此配对的 1mL 石英比色皿，微量离心管，紫外分光光度计（紫外灯已预热），离心管架。

注意：玻璃或塑料比色皿吸收紫外光，因此必须用石英比色皿测量 DNA 和 RNA 浓度。

四、实验步骤

（1）在紫外分光光度计波长 260 nm、280 nm 和 310 nm 处，将样品 DNA 溶液在 TE（或蒸馏水）中按 1 : 20 或者更高的比例稀释。用 TE（或蒸馏水）做空白校正。

（2）将各 DNA 稀释溶液装入比色皿比色，读取上述三个波长下的光密度（OD）。

（3）记录光密度数据，通过计算确定 DNA 的浓度和纯度。

对于单链 DNA 分子，其浓度（微克 / 毫升）可以通过以下公式导出：

ssDNA：33（OD260−OD310）× 稀释倍数。

对于双链 DNA 分子，其浓度（微克 / 毫升）可以通过以下公式导出：

dsDNA：50（OD260−OD310）× 稀释倍数。

对于单链 RNA 分子，其浓度（微克 / 毫升）可以通过以下公式导出：

ss RNA：40（OD260−OD310）× 稀释倍数。

注释：

① OD310 是背景吸收，盐浓度高，该值就高。

② DNA 的 OD260−OD280 值应该在 1.8 左右，高于或低于该值表示在当前 DNA 样品中存在 RNA 和蛋白。

③ RNA 的 OD260−OD310 值应该是 2.0 左右。

五、实验结果

计算所测的 DNA 浓度，并分析结果。

在 DNA 定量中，从分光光度计读取的计数是光学密度。

附录：

NanoDrop-2000 超微量分光光度计是 NanoDrop 的最新产品，近些年来不同实验室用得也越来越普遍。该仪器操作简单，几乎不需要耗材，样品体积只需要 0.5~2 μL，每样品耗时也就几秒。下面做一介绍。

精巧、简单、高通量 DNA 浓度的新仪器 NanoDrop-2000

1. 简介

NanoDrop-2000 是一款测量 DNA、RNA 和蛋白的微量分光光度计，其检测样品体积小到只需要 0.5 μL，并可得到样品浓度和纯度的比率及充分的光谱数据的报告。只需要少于 5s 的测量时间——仅仅只是移液管、测量、并擦干净的时间。特点如下：

（1）测量体积为 0.5 ～ 2 μL 的 DNA、RNA 和蛋白样品。

（2）测量的浓度高达 15 000 ng/μL，免去了稀释的步骤。

（3）低成本——不需要平板或其他耗材。

（4）对蛋白质有很好的低波长吸光度，如多肽的吸光度在 205 nm。

2. 使用方法

DsDNA：在主画面点选 Nucleic Acid，计算机与仪器自动完成联机。依照 DNA 所溶于之液体准备该溶液，取出 1.5 μL 点在侦测台上，放下上臂后再按 Blank。在右上方拉选 Sample Type 从中选 DNA-50，在 Sample ID 位置输入样品名称，将样品混匀，取出 1.5 μL 点在侦测台上，放下上臂后再按 Measure。

3. 结果整理

NanoDrop-2000 软件一开始会询问档案欲存至何处，若未指定，则档案会存在上一个使用者的文档内。

4. 注意事项

（1）侦测后立即使用拭镜纸擦拭台面。拭镜纸可折叠四次以单方向多次擦拭台面（测量 DNA 擦 5 次，测量 Protein 擦 20 次）。

（2）同一滴液体只能做一次侦测。欲重复定量同一样品，请擦掉前一滴，重新取出一滴进行侦测。

（3）核酸样品基本上可使用 1~2 μL 做测量，原则上不超过 2 μL。唯蛋白质样品因呈色剂与蛋白质本身特性，务必使用 2 μL 进行侦测。

（4）当软件跳出错误讯息时，请详细阅读帮助并依指示进行障碍排除。最常见的问题是样品内有气泡，可将上臂拉起后，擦掉该滴样品，再重新进行侦测，必要时可将样品体积加大至 2 μL。

5. 日常与定期维护

平时保持台面及仪器本身清洁。定期校正。

Experiment 21   Determination of Concentration and Purity of DNA by UV Spectroscopy

I . Experimental Objectives

By the end of this laboratory exercise, you should be able to determine the concentration and purity of DNA by UA spectroscopy.

II . Experimental Principles

The accurate quantization of concentration and relative purity of DNA is very important in the molecular manipulation, because it can often significantly improve both the quality and consistency of your experimental results no matter whether to involve the legation of purified DNA fragments into cloning vectors, the sequencing of miniprep or PCR-generated DNA, or other more specialized procedures. UV spectrophotometric determination is an accurate, rapid, and nondestructive method to determine the concentration as low as 2.5 μg/mL of DNA and is commonly used in experiments of molecular biology .

The bases on DNA (or RNA) molecules have a characteristic UV absorption peak at 260 nm. This absorption is proportional to the amount of DNA (or RNA) in the system. Using a 1cm light path, the extinction coefficient for DNA at 260 nm is 20. By using 1cm quartz cuvette, 50 ug/mL solution of double stranded DNA is equal to 1 at 260 nm. The transition between the molecules of double stranded and single stranded forms may produce certain changes on the level of absorption, but this deviation can be calibrated with a certain formula. This method is convenient and relatively accurate. The absorption peak of proteins is approximately 280 nm mainly due to tryptophan residues. The ratio of A260/A280, therefore, is a measure of the purity of a DNA preparation and should fall between 1.65 and 1.85. A lower value suggests protein contamination. If phenol, which has one maximal absorption peak at 270 nm, is contaminating the DNA preparation, then the A260 will be abnormally high, leading to an overestimation of the DNA concentration.

III . Experimental Materials

DNA sample, matching pairs of 1 mL quartz cuvettes, microcentrifuge tubes, UA spectrophotometer (UA lamp prewarmed), centrifugal tube rack.

Note: As glass or plastic curettes all absorb UA, quartz curettes must be used to measure the concentration of DNA and RNA.

IV . Experimental Procedures

(1) Make 1 : 20 or higher dilution of the stock DNA solution in TE (or ddH$_2$O). Use TE (or ddH$_2$O) as blank to calibrate the UA spectrophoto at 260 nm, 280 nm, and 310 nm wavelengths.

(2) Load each dilution of DNA solution into the cuvette and read the optical density (OD) at all the three wavelengths.

(3) Record the OD reading and determine the purity and concentration of DNA by calculation.

For ssDNA, the concentration in microgram per mL, can be derived by the following formula:

ssDNA=33(OD260−OD310) × (dilution factor)

For dsDNA, the concentration in microgram per mL, can be derived by the following formula:

dsDNA=50(OD260−OD310) × (dilution factor)

For ssRNA, the concentration in microgram per mL, can be derived by the following formula:

ssRNA=40(OD260−OD310) × (dilution factor)

Notes:

① OD310 is the absorption of background; it will be higher if the salt concentration is higher.

② OD260−OD310 for DNA should be around 1.8. Ratio above or below the number usually indicates the presence of RNA or protein concentration in the DNA sample.

③ OD260−OD310 for RNA is around 2.0.

V . Experimental Results

Measure and calculate the DNA concentration and analyze the results.

The number that you read from the spectrophotometer is the optical density or OD.

Appendix:

NanoDrop-2000(ultra micro spectrophotometer) is the latest product of NanoDrop, which has been often used in different laboratories in recent years. The instrument has the advantages of simple operation, almost no need of materials, only 0.5−2 μL volume of sample, and a few seconds of detection time for every sample. The introduction is as follows.

Smart, simple and robust NanoDrop-2000

1. Introduction

The NanoDrop-2000 is a microvolume spectrophotometer for measuring the concentration of DNA, RNA, and protein. The NanoDrop-2000 accurately measures samples as small as 0.5 μL, and reports the concentration, purity ratio, and full spectral data of a sample. The time of fast measurement is less than 5 seconds—just for pipette, measuring, and wiping clean. The features are as follows:

(1)Measure 0.5-2 μL DNA, RNA, and protein samples.

(2)Measure concentration up to 15,000 ng/ μL (dsDNA), eliminating the need for dilution.

(3)Low-cost operation−no plates or other consumables needed.

(4)Perfect for proteins with low wavelength absorbance, such as peptides at 205 nm.

2. Method of use

DsDNA: Choose the Nucleic Acid in the main screen, and then the NanoDrop-

2000 instrument will automatically connect with the computer. Use the liquid preparing to dissolve DNA, pipette 1.5 μL of this liquid on the detection station, then put down the instrument's arm and then press Blank from the screen. Select Sample Type DNA-50 at the upper right menu; input the name of the samples in the Sample ID position. Mix DNA samples, put 1.5 μL of the sample on the detection station; last, put down the arm and press the Measure.

3. Results arrangement

In the beginning NanoDrop-2000 software will ask where to save the file. If not specified, the file will be saved in the name of the last user.

4. Notes

(1) After detection, immediately use the lens paper to wipe the detection station. The table is repeatedly wiped using the lens paper with single direction which can be folded four times (DNA to 5 times, Protein to 20 times).

(2) A drop of liquid can make only one time of detection. In order to repeat the detection on the same sample, please erase the previous drop and put a new drop.

(3) Basically detection of nucleic acid samples can be done using 1−2 μL of the sample and don't exceed 2 μL in the principle. Make sure to use the 2 μL to detect the only protein samples for pigment and protein characteristics.

(4) If the error message arises out of the software, please read the troubleshooting carefully in accordance with the instructions. The most often occurring error in the detection is the sample with bubble. To resolve it, pull up the arm, wipe off the drop of the sample and add a new one to detect. If necessary, the sample volume can be enlarged to 2 μL.

5. Daily and regular maintenance

Usually keep the table and the instrument clean and regularly calibrate the instrument.

实验二十二　聚合酶链式反应

一、实验目的

学习 PCR（Polymerase Chain Reaction）反应的基本原理与实验技术，了解引物设计的一般要求。

二、实验原理

聚合酶链式反应简称 PCR，是一种分子生物学技术，用于放大扩增特定的 DNA 片段，并产生成千上万个目的序列拷贝。该方法主要依赖热循环反应，其过程包括循环加热、DNA 冷却退火和聚合酶催化等一系列合成反应。引物序列（即 DNA 小片段）与目标区域互补，与 DNA 聚合酶一起进行选择和重复扩增。随着 PCR 进行，以单链 DNA 为模板进行 DNA 复制和延伸。PCR 反应可以通过扩展改良，可以开展更为广泛的遗传操作实验。我们编制的热循环反应基于 3 个不同的温度。首先是变性温度，为 95℃；其次是较低的复性温度，为 37℃~55℃，此时引物和模板进行退火；最后是合成延伸温度，为 72℃。循环数依据需要为 20~35，直到扩增产生足够我们需要的目标 DNA。

三、实验材料

PCR 仪、离心机、移液枪、制冰机、电泳设备、核酸紫外检测仪、0.2 mL PCR 反应管、移液枪枪头。

10×PCR 反应缓冲液，25 mmol/L $MgCl_2$、10 mm dNTP，Taq DNA 聚合酶（5U/μL），50 μmol/L 引物 1 和 2，模板 DNA，灭菌水，矿物油（如果 PCR 仪没有热盖）。

四、实验步骤

（1）准备 PCR 反应溶液。

取下表所示混样放入 0.2mL 的反应管中。

成分	需要体积 / μL
DNA 模板	1
前引物 (0.5 mg/ mL)	0.5
后引物 (0.5 mg/ mL)	0.5
dNTP (1.25 mmol/L each)	1
10 × PCR 反应缓冲液	1
聚合酶 Taqpolymerase (2U/ mL)	0.2
灭菌水	6

（2）准备一个不加模板的反应管为对照。

（3）如果 PCR 仪没热盖，每个反应需要加两滴石蜡油。

（4）离心管预热 94℃。

（5）按下列参数循环运行。

循环次数	参数
1	94℃，60 s
25	94℃，45 s
	57℃，45 s
	72℃，1 min
1	72℃，7 min

注意：

这是一个非常保守的程序，在大多数情况下都能扩增出条带，但容易产生非特异性条带。PCR 反应可以按比例缩小反应体积到 25 μL。进一步缩小需要矿物油覆盖以防止在热循环过度蒸发。用阴性对照检查是否污染是必要的。阴性对照组各反应成分没有模板。阳性对照也可以用来防范扩增误差。

五、实验结果

思考题：

PCR 的原理是什么？

Experiment 22　The Polymerase Chain Reaction (PCR)

Ⅰ. Experimental Objectives

Learn the basic principle and the experimental technology of PCR reaction and understand the general requirements for primer design.

Ⅱ. Experimental Principles

The polymerase chain reaction (PCR) is a biochemical technology in molecular biology of amplifying a single or a few copies of a piece of DNA across several orders of magnitude, generating thousands to millions of copies of a particular DNA sequence. The method relies on thermal cycling, consisting of cycles of repeated heating and cooling of the reaction for DNA melting and enzymatic replication of the DNA. The primers (short DNA fragments) with sequences complementary to the target region enable selective and repeated amplification along DNA with a polymerase. As PCR progresses, the DNA generated itself is used as a template for replication. PCR can be extensively modified to perform a wide array of genetic manipulations. We program the thermal cycler at three different temperatures. First a high temperature (about 95℃) is used to separate the DNA strands. Second a relatively low temperature (about 37℃ −55℃) to allow the primers to anneal to the template DNA STRANDS. At last a medium temperature (about 72℃) to allow DNA synthesis. The number of cycle is usually about 20-35 until we produce as much as amplified DNA we need.

III . Experimental Materials

PCR equipment, centrifuges, pipette, ice machine, electrophoresis equipment, nucleic acid ultraviolet detector, 0.2mL PCR reaction tube and pipette tip top.

10 × amplification buffer; 25 mmol/L $MgCl_2$; 10 mm dNTP; 5 unit/μL Taq DNA Polymerase; 50 μmol/L oligonucleotide primer 1 and 50 μmol/L oligonucleotide primer 2; template DNA (1 μg genomic DNA); sterile water; mineral oil (for thermocyclers without a heated lid).

IV . Experimental Procedures

(1) Prepare the standard PCR reaction mix.

Put the following items shown in the table in a 0.2 mL tube.

Gradient	Add volume / μL
DNA template	1
Forward primer (0.5 mg/ mL)	0.5
Reverse primer (0.5 mg/ mL)	0.5
dNTP (1.25 mmol/L each)	1
10 × buffer	1
Taqpolymerase (2U/ mL)	0.2
ddH$_2$O	6

(2) Prepare a control reaction tube with no template DNA.

(3) If the thermocycler does not have a heated lid, add 2 drops of silicone oil to each reaction.

(4) Place tubes in a thermal cycler preheated to 94℃ .

(5) Run the following program:

Cycling	Parameters
1	94℃，60 s
25	94℃，45 s
	57℃，45 s
	72℃，1 min
1	72℃，7 min

Note:

This is an extremely conservative protocol designed to give bands under most circumstances. It is prone to give non-specific bands. I've successfully scaled down the reaction volume to 25 μL. Further scaling down requires an overlay of mineral oil to prevent excessive evaporation during cycling. A negative control is necessary to check for contamination. Negative controls have all the reaction components except for the template. A positive control can also be included to guard against the error of PCR.

Ⅴ. Experimental Results

Question:

What is the principles of PCR?

实验二十三　用琼脂糖凝胶电泳法检测 DNA

一、实验目的

检测目的 PCR 产物和其量化的大小（DNA 分子的长度）。

二、实验原理

凝胶电泳是一种广泛用于核酸和蛋白质的分析技术，在分子生物学研究中大多数实验室经常利用琼脂糖凝胶电泳分析。凝胶电泳法是一种在电场作用下基于移动速率来分离物质的方法。琼脂糖是从海藻多糖中纯化得来的。琼脂糖凝胶制备方法是将干琼脂糖悬浮在缓冲溶液里，煮沸至溶液变澄清，然后将其倒入一个槽中冷却，最终形成明胶板。在电泳时将凝胶浸入含有正负电极室的缓冲溶液中，使 DNA 在电场作用下移动。影响 DNA 迁移速度的因素包括电场强度、琼脂糖凝胶浓度以及长度不同的 DNA 分子等，小分子的 DNA 比大分子 DNA 移动快。DNA 本身在凝胶中无法看到，但可以通过将 DNA 和染料结合来观察 DNA。溴化乙啶可插入 DNA 分子的双链中。在紫外光的照射下，插入溴化乙啶的 DNA 呈橙红色荧光，所以溴化乙啶可以作荧光指示剂指示 DNA 的含量和位置。

三、实验材料

琼脂糖、Eppendorf 管、Tip 头、TAE 电泳缓冲液、6× 上样缓冲液、DNA 标样、电泳槽、电源、染色盘、手套、微量进样器、DNA 染料。

四、实验步骤

1. 凝胶准备

（1）称取琼脂糖 1.25 g，放入 500 mL 的三角瓶中。加入 125 mL TAE 溶液到三角瓶中（总的凝胶的体积取决于托盘的尺寸）。

（2）置微波炉中或水浴锅加热，至溶液变得清亮（如果用微波炉要多次加热，以防止溢出）。

（3）冷却溶液到 50℃ ~55℃，不时摇动三角瓶。

（4）用透明胶布密封电泳槽的两端。将梳子插入槽中。

（5）将琼脂糖倒入到槽中，自然冷却（凝固后变成奶白色）。小心去掉梳子和透明胶布。

（6）将其放入电泳槽，加入适量的 TAE 缓冲液，液面高于胶面 2~3 nm。

注：凝胶可以使用前几天做好，并密封在塑料包装中（没有梳子）。如果凝胶变得过于干燥，加载样品之前使其在缓冲液中吸水几分钟。

2. 上样

（1）待测的 DNA 样品中，加 2 μL 溴酚蓝指示点样缓冲液。

（2）记录样品点样秩序与点样量。包括准备样品，DNA 模板，不同生物体的 DNA，对照以及标样。

（3）每个胶孔仔细加混样 10 μL。

（4）至少在每一排上的凝胶加 5 μL 的 DNA 标样。

注：如果电泳多个凝胶，可通过在不同泳道加载的 DNA 标样来避免样品混乱。

3. 电泳

（1）盖上盖子，开启电源开关。

（2）连接电源，确保正级（红色）和负级（黑色）正确连接。

（3）打开电源，电压 100V。最大允许电压会根据电泳槽的大小而变化。

（4）通过在每个电极上形成的气泡检查以确保当前正电源正在运行。

（5）通过观察蓝色染料运动方向——这将需要几分钟，确保电流是在正确的方向运行（蓝色染料将与 DNA 在同一方向运行）。

（6）当蓝色染料运行到凝胶末端，关掉电源。

（7）断开电源，除去电泳室盖。使用手套，小心地取出托盘和凝胶。

4. 染色

（1）戴上手套，从托盘取出凝胶并放入染色盘中。

（2）添加染色液混合。在添加 0.5 μg/mL 溴化乙啶溶液中染色 10~15 min。

（3）倒出染色（染色可以保存以供将来使用）。电泳凝胶块直接在紫外灯下拍照或绘图，通过已知的分子标记条带来估计扩增条带大小。

五、实验结果

绘制电泳图谱，并对结果进行分析。

思考题：

有哪些因素会影响到电泳图谱上 DNA 条带的位置？

附录：

TAE 缓冲液配方：

　　4.84 g Tris Base

　　1.14 mL 冰醋酸

　　2 mL 0.5M EDTA（pH 8.0）

　　加水使总体积为 1 L

配制方法：加 Tris base 到 900 mL 水中，然后添加乙酸和 EDTA 并混合。将混液倒入 1 L 量筒，添加水使总体积为 1 L。为方便使用通常配 TAE 缓冲（10 倍或 50 倍），使用时提前用水稀释 1 倍。

6× 上样缓冲液配方：

　　1 mL 无菌水

　　加入 1 mL 甘油

　　加入溴酚蓝（~0.05 mg）

　　为长期存储，保持样冷冻。

QUIKView DNA 染色配方：

　　25 mL QUIKView DNA 染色

　　加 475 mL 水（50℃~55℃）

制备琼脂糖凝胶

按照被分离 DNA 分子的大小，决定凝胶中琼脂糖的百分含量。一般情况下，可参考下表：

项　目	不同琼脂糖凝胶浓度		
	0.7%	**1.0%**	**2.0%**
琼脂糖 /g	1.05	1.5	3.0
20 倍 TAE/mL	7.5	7.5	7.5
ddH$_2$O/mL	142.5	142.5	142.5
EtBr (5mg/mL)/ μL	25	25	25
总体积 /mL	150	150	150

Experiment 23   Detection of DNA by Agarose Gel Electrophoresis

Ⅰ. Experimental Objectives

Determine the presence or absence of PCR products and quantify the size (length of the DNA molecule) of the products.

Ⅱ. Experimental Principles

Gel electrophoresis is a widely used technique for the analysis of nucleic acids and proteins. Most laboratories of molecular biology routinely use agarose gel electrophoresis for the preparation and analysis of DNA. Electrophoresis is a method of separating substances based on the rate of movement, under the action of an electric field. Agarose is a polysaccharide purified from seaweed. An agarose gel is created according to the following steps of suspending dry agarose in a buffer solution, boiling until the solution becomes clear, and then pouring it into a casting tray and allowing it to cool. The result is a flexible gelatin-like slab. Before running the gel electrophoresis, the agarose gel is submersed in a chamber containing a buffer solution with a positive and negative electrode. Under the action of an electrical field, DNA will move to the positive electrode (red) and away from the negative electrode (black). Several factors influence how fast the DNA moves, which are the strength of the electrical field, the concentration of agarose in the gel and most importantly the sizes of the DNA molecules. Smaller molecules of DNA move through the agarose faster than the larger molecules. DNA itself is not visible within an agarose gel. The DNA will be visualized by the use of a dye that binds to DNA. EB can be inserted into a double-stranded DNA molecule. Under UV irradiation, the DNA inserted ethidium bromide shows the orange of fluorescence so it can be used as indicator for DNA content and location.

III . Experimental Materials

Agarose, Eppendorf tube, Tip, TAE buffer, $6 \times$ sample loading buffer, DNA ladder standard, electrophoresis chamber, power supply, staining tray , gloves, pipette and tips, QUIKView DNA Stain.

IV . Experimental Procedures

1. Preparing the agarose gel

(1) Measure 1.25 g agarose powder and add it to a 500 mL flask, then pour 125 mL TAE buffer into the flask (the total gel volume well varies depending on the size of the casting tray).

(2) Melt the agarose in a microwave or hot water bath until the solution of agarose becomes clear (If melt in a microwave, the solution must be heated for several short intervals—do not let the solution boil out of the flask).

(3) Let the solution cool to 50℃ −55℃ , then swirl the flask occasionally to cool evenly.

(4) Seal both the ends of the electrophoresis tank with transparent tape. Insert the combs into the gel casting tray.

(5) Pour the melted agarose solution into the casting tray and let it cool naturally (it should appear milky white). Carefully pull out the combs and remove the tape.

(6) Place the gel in the electrophoresis chamber. Add enough TAE buffer so that there is a 2−3 nm distance over the gel.

Note: Gels can be made several days prior to its using and sealed in plastic wrap (without combs). If the gel becomes excessively dry, allow it to rehydrate in the buffer within the gel box for a few minutes prior to loading samples.

2. Loading the gel

(1) Pipette 2 μL of sample loading buffer to each 10 μL PCR reaction.

(2) Record the order in which each sample will be loaded on the gel, and prepare the sample, the DNA template, what organism the DNA came from, controls and ladder.

(3) Carefully pipette 10 μL of each sample/Sample then load the buffer mixture into separate wells in the gel.

(4) Pipette at least 5 μL of the DNA ladder standard into one well for each row on the gel.

Note: if you are running multiple gels, load the DNA ladder in different lanes on each gel to avoid later confusion.

3. Running the gel

(1) Cover the gel box with the lid, and then connect the electrodes.

(2) Connect the electrode wires to the power supply, and make sure that the positive (red) and negative (black) are correctly connected.

(3) Turn on the power supply to about 100 volts. Maximum voltage allowed will vary depending on the size of the electrophoresis chamber.

(4) Check the bubbles forming on each electrode to make sure that the current is running.

(5) Check the movement of the blue loading dye to make sure that the current is running in the correct direction—this will take a couple of minutes (it will run in the same direction as the DNA).

(6) Let the power run until the blue dye approaches the end of the gel. Turn off the power.

(7) Disconnect the wires from the power supply. Remove the lid of the electrophoresis chamber. Carefully remove the tray and gel with gloves.

4. Gel Staining

(1) Remove the gel from the casting tray with gloves, and put it in the staining dish.

(2) Pour dye liquid into the dish and mix. Allow the gel to stain for 10-15 minutes

with 0.5 μg/mL ethidium bromid solution.

(3) Pour off the dye liquid. View the gel under the UV light, photograph and estimate the size of the product bands based on the bands from the known molecular-weight markers.

V . Experimental Results

Draw the map of the electrophoresis, and analyze the results of electrophoresis.

Question:

What kinds of factors will affect the DNA strip position of the electrophoresis?

Appendix:

Ingredients of TAE buffer

　　4. 84 g Tris Base

　　1.14 mL Glacial Acetic Acid

　　2 mL 0.5 mol/L EDTA (pH 8.0)

　　Bring the total volume up to 1L with water

Preparation method: Add Tris Base into 900 mL H_2O, then add acetic acid and EDTA to solution and mix. Pour the mixture into 1 L graduated cylinder and add H_2O to a total volume of 1 L. For convenient use, a stock of TAE buffer (either 10 × or 50 ×) is often prepared and diluted with water to 1 × concentration prior to its using.

6× sample loading buffer

　　1 mL sterile H_2O

　　1 mL Glycerol

　　Add enough bromophenol blue to make the buffer deep blue (~ 0.05 mg).

　　For long term storage, keep the sample loading buffer in the fridge.

QUIKView DNA Stain

　　25 mL QUIKView DNA Stain

　　Add 475 mL water (50℃ −55℃)

Preparation of the gel

Determine the percentage of agarose gel in accordance with the molecular size of the isolated DNA. Under normal circumstances, you can refer to the following table:

Items	Different Percentages of agarose		
	0.7%	1.0%	2.0%
Agarose/g	1.05	1.5	3.0
$20 \times$ TAE/mL	7.5	7.5	7.5
ddH$_2$O/mL	142.5	142.5	142.5
EtBr (5mg/mL)/ μL	25	25	25
Total vol/mL	150	150	150

附：实验报告格式

实验报告

实验课程：＿＿＿＿＿＿＿＿＿＿＿＿＿＿

学生姓名：＿＿＿＿＿＿＿＿＿＿＿＿＿＿

学　　号：＿＿＿＿＿＿＿＿＿＿＿＿＿＿

专业和班级：＿＿＿＿＿＿＿＿＿＿＿＿＿

年　　月　　日

实验预习报告

（没有预习不能参加实验）

实验题目：_____

学生姓名：_____　　学　　号：_____　　专业和班级：_____

实验类型：□验证　　□综合　　□设计　　□创新

实验日期：_____　　实验成绩：_____　　教师签名：_____

1. 实验目的

2. 实验内容

3. 实验注意事项

4. 主要实验步骤

实验报告

实验题目：_____

学生姓名：_____　　学　　号：_____　　专业和班级：_____

实验类型：□验证　　□综合　　□设计　　□创新

实验日期：_____　　实验成绩：_____　　　教师签名：_____

1. 实验目的

2. 实验材料

3. 实验内容

4. 结果分析

5. 讨论

6. 对改进实验的建议

实验报告

（综合性实验）

实验题目：＿＿＿＿＿＿＿＿＿＿＿＿＿＿＿＿＿＿＿＿＿

学生姓名：＿＿＿＿＿＿　学　　号：＿＿＿＿＿＿　专业和班级：＿＿＿＿＿

实验类型：综合＿＿＿＿＿＿＿

实验日期：＿＿＿＿＿＿　实验成绩：＿＿＿＿＿＿　教师签名：＿＿＿＿＿

1. 摘要

2. 关键词

3. 引言

4. 材料与方法

5. 结果与分析

6. 讨论

7. 参考文献

实验报告

(设计性实验)

实验题目: _____

学生姓名: _____　　学　　号: _____　　专业和班级: _____

实验类型: 设计

实验日期: _____　　实验成绩: _____　　教师签名: _____

1. 摘要

2. 关键词

3. 实验原理

4. 仪器和试剂

5. 样品制备和测定

6. 结果分析与讨论

7. 参考文献

Appendix: The Format of the Laboratory Report in English

Laboratory Report

Experiment course _____

Student name _____

Student ID _____

Major and Class _____

Year   Month   Date

Experiment Previewing Report

(NO one can participate in the experiment without preview)

Experimental subject: _____

Student name: _____ Student ID: _____ Major and class: _____

Experiment type: ☐ verification ☐ comprehensive ☐ design ☐ innovation

The date of the experiment: _____ Score: _____ Teacher signature: _____

1. Experimental Objectives

2. Experimental contents

3. Matters needing attention

4. Main experimental procedures

Experiment Report

Experimental subject: _____

Student name: _____　　Student ID: _____　　Major and class: _____

Experiment type: □ verification　□ comprehensive　□ design　□ innovation

The date of the experiment: _____　　Score: _____　　Teacher signature: _____

1. Experimental Objectives

2. Experimental Materials

3. Experimental contents

4. Results analyses

5. Discussions

6. Suggestions to improve the experiment

Experiment Report

(Comprehensive experiment)

Experimental subject: _____

Student name: _____   Student ID: _____   Major and class: _____

Experiment type: ☐ comprehensive

The date of the experiment: _____   Score: _____   Teacher signature: _____

1. Abstract

2. Keywords

3. Introduction

4. Materials and methods

5. Results and analyses

6. Discussions

7. References

Experiment Report

(*Design experiment*)

Experimental subject: _____

Student name: _____　　Student ID: _____　　Major and class: _____

Experiment type: □ design

The date of the experiment: _____　　Score: _____　　Teacher signature: _____

1. Abstract

2. Keywords

3. Experimental principles

4. Instruments and reagents

5. Examination and preparation of samples

6. Results analyses and discussions

7. References

参 考 文 献

［1］刘祖洞，江绍慧．遗传学实验（第二版）［M］．北京：高等教育出版社，1987．

［2］季道潘．浙江农业大学·遗传学实验［M］．北京：中国农业出版社，1994．

［3］姜泊，张亚历，周殿元．分子生物学常用实验方法［M］．北京：人民军医出版社，1996．

［4］王亚馥，戴灼华．遗传学［M］．北京：高等教育出版社，1999．

［5］孙勇如．遗传学手册［M］．长沙：湖南科学技术出版社，1989．

［6］徐秀芳，张丽敏，丁海燕．遗传学实验指导［M］．武汉：华中科技大学出版社，2013．

［7］焉慧民，袁文静．遗传学实验［M］．武汉：武汉大学出版社，1994．

［8］张文霞，戴灼华．遗传学实验指导［M］．北京：高等教育出版社，2007．

［9］祝水金．遗传学实验指导（第二版）［M］．北京：中国农业出版社，2005．

［10］Birren, Green E D, Klapholz S, Myers R M and Roskams J. 1997. Analyzing DNA：A Laboratory Manual［M］. New York：Cold Spring Harbor Laboratory Press.

［11］Robert F, Weaver. 2002. Molecular Biology［M］. Second Edition. New York：M cG raw-Hill Companies, Inc.